Contemporary's

NUMBER POWER

Problem Solving and Test-Taking Strategies

7

ELLEN C. FRECHETTE

Project Editor:
Caren Van Slyke

CONTEMPORARY
BOOKS
CHICAGO

Published by Contemporary Books, Inc.
Two Prudential Plaza, Chicago, Illinois 60601-6790
Manufactured in the United States of America
International Standard Book Number: 0-8092-4195-1

Published simultaneously in Canada by
Fitzhenry & Whiteside
195 Allstate Parkway
Markham, Ontario L3R 4T8
Canada

CONTENTS

(continued)

Chapter 5: Making a Plan

Chapter 6: Solving the Problem

Chapter 7: Checking the Answer

Chapter 8: Using Charts, Graphs, and Drawings

Chapter 9: Working Geometry Word Problems

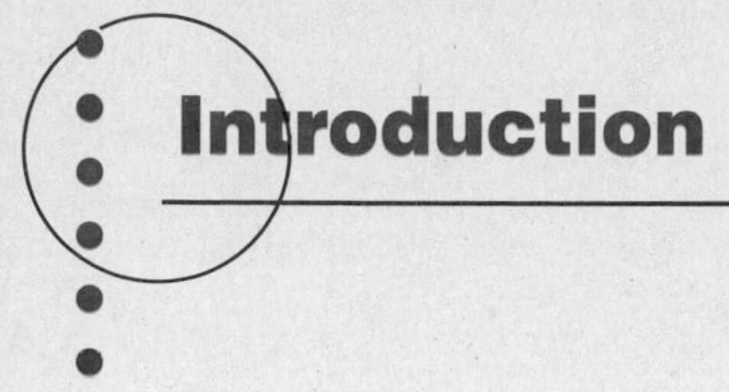

Introduction

To succeed in math classes and on math tests, students need to become effective problem solvers. Even if they master basic math skills, many students still find it a challenge to interpret word problems, charts, graphs, and diagrams.

Number Power 7: Problem Solving and Test-Taking Strategies gives students the tools they need to understand and solve a wide variety of problems. This text targets the **comprehension**, **organizational**, and **reasoning** skills required to pass the GED Math Test, state-wide high school proficiency exams, and many other standardized tests.

This book is organized so that each problem solving strategy can be applied to different levels of math skills.

- A sample problem is presented at the beginning of each lesson to introduce each new idea. The math on the sample problem is **generally** restricted to whole numbers, money problems, or simple measurements. This will ensure that most students will be able to grasp the problem solving concept while they are still developing their computational skills.
- After the problem solving strategy has been introduced, students will apply it to problems. Since students will be at differing levels, each practice problem will be accompanied by a symbol based on its highest skill level:

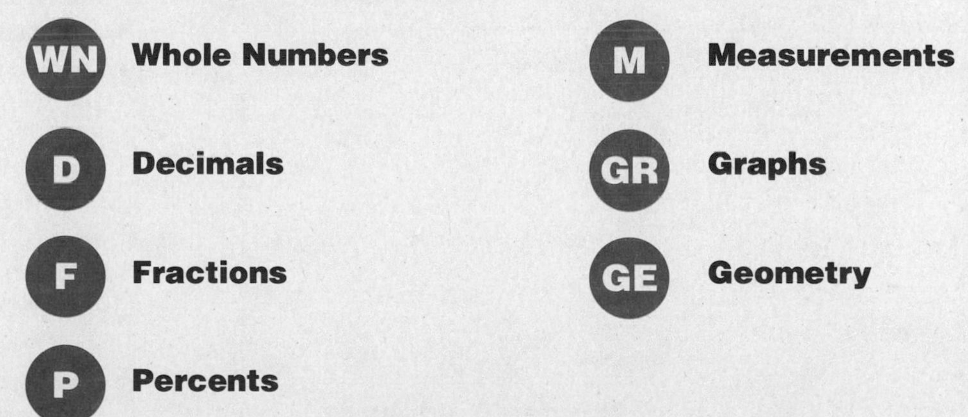

Depending on students' computational backgrounds, they will be able to do some, many, or all the problems in a practice set.

However students use *Number Power 7,* they will find a wealth of practical techniques to build reasoning and analytical skills that can be used throughout their lives.

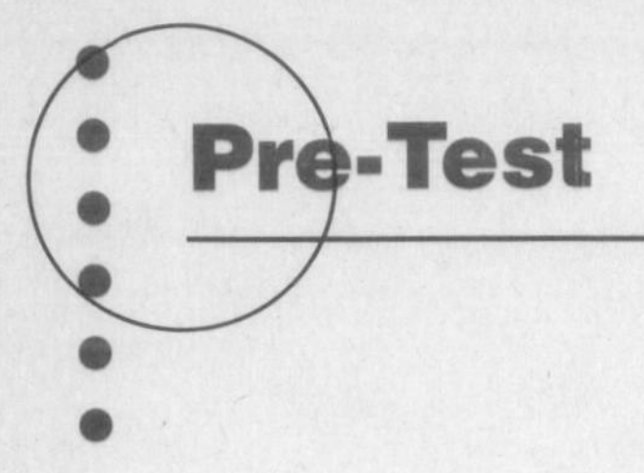

Pre-Test

Directions: For each problem, choose the best answer from the choices given.

1. If Tammy earns $47.90 in one 4-hour shift, how much does she earn per hour?

Which of the following is a correct restatement of the question in the problem above?

(1) Find Tammy's pay per day.
(2) Find how much Tammy earns during each shift.
(3) Find Tammy's wages before taxes.
(4) Find how much Tammy earns in one year.
(5) Find how much Tammy is paid for one hour.

2. Carmine drives 6 miles to work each day and the same distance home again. If he worked 4 days this week, how many miles did he drive in all for work?

(1) 6
(2) 10
(3) 12
(4) 24
(5) 48

3. According to the chart, how many pounds can a 3-cubic-foot crate hold?

(1) 60
(2) 100
(3) 140
(4) 180
(5) 200

Packing Guidelines

Crate	Weight
1 cu ft	60 lb
2 cu ft	100 lb
3 cu ft	140 lb
4 cu ft	200 lb

4. Kay planted 6 rows of lettuce with 6 plants in each row. Each row was 6 feet long, and it took her 6 hours to do all the planting. How many lettuce plants did she plant in all?

The following information is **not** needed to solve the problem:

(1) 6 feet, 6 hours
(2) 6 plants
(3) 6 feet
(4) 6 rows, 6 plants
(5) 6 rows

5. Max spent 18 hours assembling 3 picnic tables. On average, how many hours did each table take to assemble?

Which of the following operations should you use to solve the problem above?

(1) Multiply 18 by 3.
(2) Divide 18 by 3.
(3) Subtract 3 from 18.
(4) Add 3 and 18.
(5) Divide 3 by 18.

6. The Halt Hunger Telethon raised $300,000 from about 10,000 callers. Approximately how much money did each caller pledge?

Without doing any calculations, choose the most sensible answer from the list below.

(1) $10
(2) $30
(3) $3,000
(4) $300,000
(5) $3,000,000

7. According to the chart, how many *yards* of twine will the packer need to wrap three #2 cartons?

(1) 3
(2) 4
(3) 5
(4) 6
(5) 12

Twine Lengths

Carton	Number of feet
#1	2
#2	4
#3	6
#4	8

8. How many **cups** of liquid are held in the containers shown?

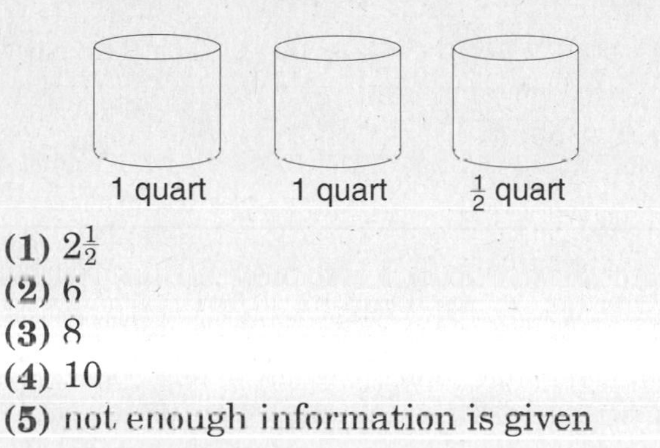

(1) $2\frac{1}{2}$
(2) 6
(3) 8
(4) 10
(5) not enough information is given

9. Lisa paid $55 for a dress at Discount Dan's and $90 for a dress at Mrs. Mooney's Shop. Which of the following expressions shows how much more Lisa paid for the more expensive dress?

(1) $90 + 55$
(2) $55 - 90$
(3) $90 - 55$
(4) $90 \div 55$
(5) 90×55

10. A mother baked 70 brownies and gave 12 of them to each of her three neighbors. Which of the following expressions shows how many brownies she had left?

(1) $70 - (3 \times 12)$
(2) $70 - 12$
(3) $(70 - 12) \times 3$
(4) $(70 - 12) \div 3$
(5) $70 - 12 - 3$

11. Which of the following expressions shows the average weight in pounds of the Sammarco children at birth, according to the chart?

Sammarco Family

Jennifer	—	6 lb
Meghan	—	7.5 lb
Jonathan	—	7.5 lb
Matthew	—	8 lb

(1) $6 + 7.5 + 7.5 + 8$
(2) $\dfrac{6 + 7.5 + 7.5 + 8}{3}$
(3) $\dfrac{6 + 7.5 + 7.5 + 8}{4}$
(4) $3 \times (6 + 7.5 + 7.5 + 8)$
(5) $4 \times (6 + 7.5 + 7.5 + 8)$

12. There are twenty-eight people on Floyd's shift at the factory. Half of them are women. How many women are on Floyd's shift?

(1) 2
(2) 14
(3) 28
(4) 56
(5) not enough information is given

13. A manufacturing plant turns out 9,400 circuit boards per shift. If there are 10 assembly lines running on each shift, with 5 employees per line, how many boards does each assembly line turn out per shift?

(1) 188
(2) 940
(3) 1,880
(4) 47,000
(5) 94,000

Questions 14 and 15 refer to the following graph.

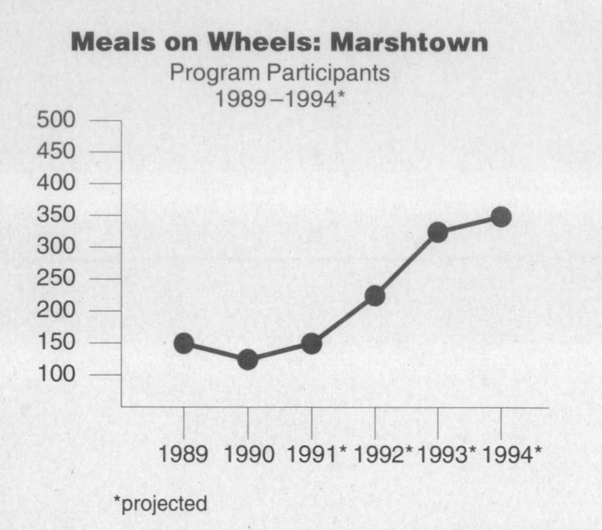

14. According to the graph, about how many more people in Marshtown are projected to use the Meals on Wheels in 1991 than in 1990?

(1) 25
(2) 120
(3) 125
(4) 150
(5) not enough information is given

15. According to the graph, how many people over the age of 65 are expected to use the Meals on Wheels program in 1993?

(1) 350
(2) 325
(3) 300
(4) 225
(5) not enough information is given

---

16. Mary and Lewis were married in 1988. Three years later they had their first baby. What more do you need to know to find out how old Mary was when she had her first child?

(1) Mary's age when she got married
(2) Lewis's present age
(3) the birth date of Mary's child
(4) the present age of Mary's child
(5) the year the baby was born

Questions 17–19 refer to the following information.

To landscape their yard, Maureen and David bought 4 cubic yards of loam at $100 per cubic yard. From the same landscaping company they purchased 4 evergreen bushes at $15.99 each and a dozen smaller shrubs at $7.98 each.

To have the landscaping company spread the loam and plant the shrubs and bushes, it would cost Maureen and David $35 per hour for labor.

17. How much money did Maureen and David spend on the bushes and shrubs?

(1) $63.96
(2) $95.76
(3) $109.72
(4) $159.72
(5) not enough information is given

18. The landscaping company estimates that it would take 22 hours to spread the loam and plant the shrubs. How much would the company charge for the labor?

(1) $35.00
(2) $720.00
(3) $770.00
(4) $879.72
(5) $929.72

19. Which expression shows how much the landscaping company would charge to do the whole job—including materials and 22 hours of labor?

(1) 22 × $35
(2) 4($100 + $15.99) + (22 × $35)
(3) (4 + $100) + (4 + $15.99) + (12 + $7.98) + (22 + $35)
(4) (4 × $100) + (4 × $15.99) + (12 × $7.98) + (22 × $35)
(5) (4 × $100) + (4 × $15.99) + $7.98 + (22 × $35)

20. Peter bought a total of 13 baseball game tickets and gave some to Sondra. He found that he had 9 left. Which of the following equations will tell you how many tickets Peter gave Sondra?

(1) $13 + 9 = x$
(2) $9 + 13 = x$
(3) $9 - x = 13$
(4) $x + 13 = 9$
(5) $13 - x = 9$

21. To make the figure shown, Su Lin needs five tablespoons of rubber cement. How many tablespoons of rubber cement would she need for a similar figure that measures 84 inches around?

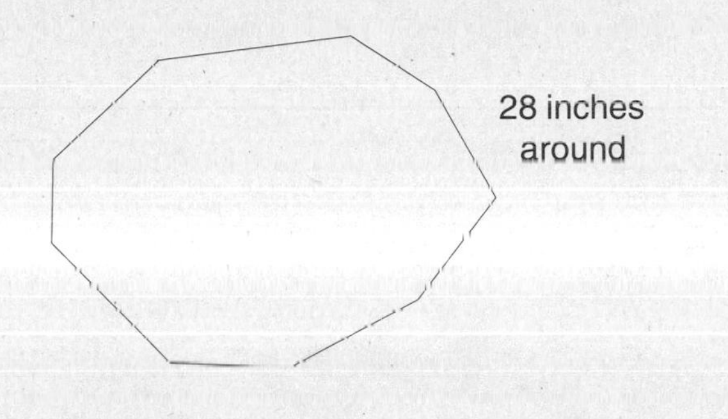

(1) 5
(2) 15
(3) 56
(4) 112
(5) 127

22. Approximately how many loads of laundry do the Mosers do each year if they wash 5 loads per week?

(1) 35
(2) 50
(3) 250
(4) 500
(5) 2,500

23. The chart shows the amount of time different participants took to complete a puzzle for a scientific research project. Which of the following shows the correct order from shortest to longest time?

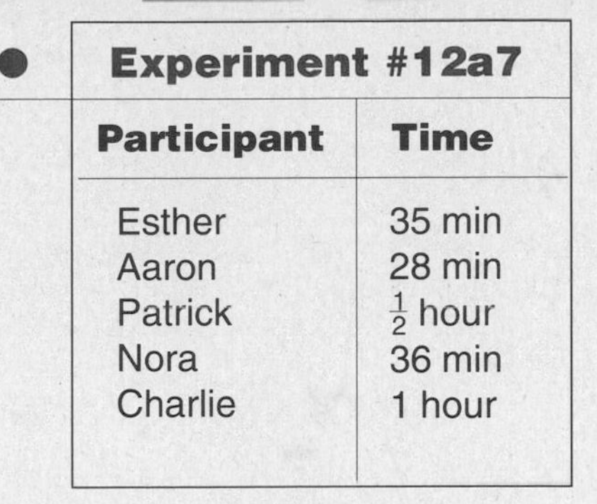

Experiment #12a7

Participant	Time
Esther	35 min
Aaron	28 min
Patrick	$\frac{1}{2}$ hour
Nora	36 min
Charlie	1 hour

(1) Aaron, Patrick, Esther, Nora, Charlie
(2) Charlie, Nora, Esther, Patrick, Aaron
(3) Aaron, Esther, Nora, Patrick, Charlie
(4) Esther, Nora, Patrick, Charlie, Aaron
(5) not enough information is given

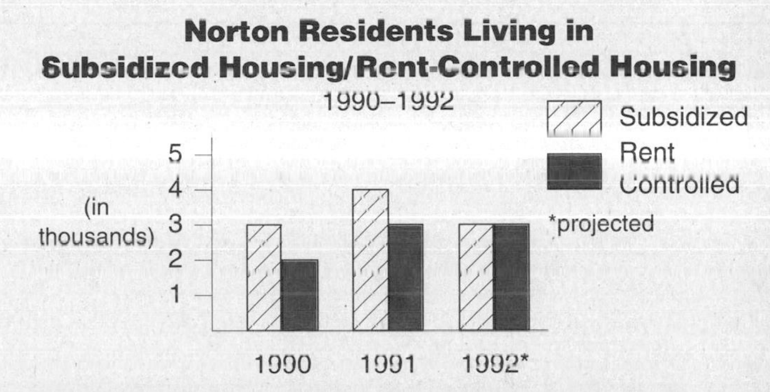

24. According to the graph, how many more Norton residents lived in subsidized housing in 1991 than in 1990?

(1) 1
(2) 4
(3) 7
(4) 1,000
(5) 4,000

25. Which of the following formulas can be used to find the number of square units in the figure?

(1) $P = a + b + c$
(2) $P = \frac{1}{2}(a + b + c)$
(3) $A = a \times b \times c$
(4) $A = \frac{1}{2}(a \times b \times c)$
(5) $A = \frac{1}{2}ab$

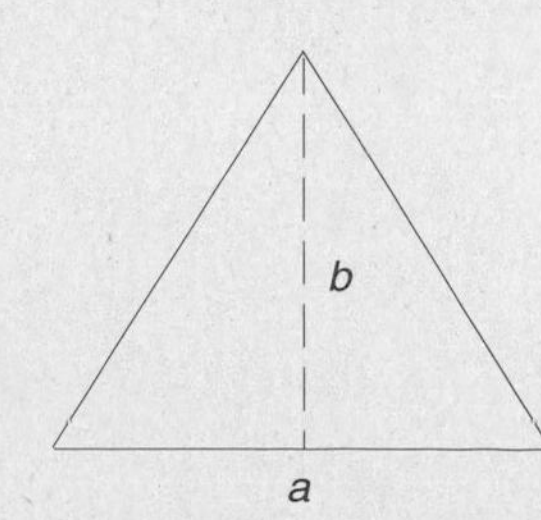

PRE-TEST DIAGNOSTIC CHART

Directions: Circle any item number that was answered *incorrectly.* This pre-test is a preview of **some** of the problem solving strategies in this book. To effectively master problem solving and test-taking strategies, you should work through the entire book. As you do, you should focus on the skills for the items that you circled.

ITEM NUMBER	SKILL AREA	CHAPTER NUMBER
2	Multi-Step Problems	1
1,7,8	Understanding the Question	3
9,10,11,19	Set-Up Questions	3,6
16	What More Do You Need to Know?	3
12	Finding "Hidden" Information	4
4,13	Extra Information	4
15	Not Enough Information	4
17,18	Working with Item Sets	4
5	Making a Plan	5
20	Writing Equations for Word Problems	5
21	Writing Proportions for Word Problems	5
14,22	When an Estimate Is the Answer	6
23	Comparing and Ordering	6
6	Is the Answer Reasonable?	7
3,24	Using Charts, Graphs, and Drawings	8
25	Let Formulas Work for You	9

PRE-TEST ANSWERS

1. (5) Find how much Tammy is paid for one hour.

2. (5) 48

6 miles	12 miles per day
× 2 trips per day	× 4 days
12 miles	48 total miles

3. (3) 140
Find 3 cubic feet on the chart; read across to the value 140 lb.

4. (1) 6 feet, 6 hours
You need to know the number of rows and the number of plants per row. The number of feet per row and the hours it took Kay to plant are unnecessary information.

5. (2) Divide 18 by 3.
To find the time it took to build each table, you need to divide the total time by the number of tables.

6. (2) $30
You can estimate $300,000 divided by 10,000; $30 is a reasonable amount of money for an average donation.

7. (2) 4

4 feet
× 3 cartons
12 total feet

3) 12 feet = 4 yards
(3 = ft per yard)

8. (4) 10 1 quart = 4 cups
$2\frac{1}{2}$ qts × 4 cups = 10 cups

9. (3) 90 – 55
more expensive dress – less expensive dress = difference

10. (1) 70 – (3 × 12)
total brownies – (number of neighbors × brownies per person)

11. (3) $\frac{6 + 7.5 + 7.5 + 8}{4}$
total weight of children ÷ number of children = average weight

12. (2) 14
half of 28 = 28 ÷ 2 = 14

13. (2) 940
10 lines) 9400 circuit boards = 940 boards per shift
(The number of employees per line is unnecessary information.)

14. (1) 25
150 (estimated 1991 value)
– 125 (1990 value)
25

15. (5) not enough information is given
The question asks for the number of Meals on Wheels participants over the age of 65; the graph does not give any information about age.

16. (1) Mary's age when she got married
If you knew Mary's age when she got married, you could add three years to that and find the age she was when she had her first child.

17. (4) $159.72

$15.99	$7.98	63.96 (bushes)
× 4	× 12	+ 95.76 (shrubs)
63.96	95.76	$ 159.72

18. (3) $770
22 hours
× $35 per hour
$770

19. (4) (4 × $100) + (4 + $15.99) + (12 × $7.98) + (22 × $35)

20. (5) 13 – *x* = 9
tickets Peter started with – number given away = number left

21. (2) 15

$$\frac{\text{tablespoons rubber cement}}{\text{inches of cardboard}} = \frac{5}{28} = \frac{x}{84}$$

$5 \times 84 = 28 \times x$
$420 = 28 \times x$
$420 \div 28 = x$
$15 = x$

22. (3) 250
There are 52 weeks in a year; round off to 50.
50 weeks × 5 loads of laundry = 250 loads

23. (1) Aaron, Patrick, Esther, Nora, Charlie

Convert times to minutes:	Esther	35 min
	Aaron	28 min
	Patrick	$\frac{1}{2}$ hr = 30 min
	Nora	36 min
	Charlie	1 hr = 60 min

Put in order of shortest to longest:	Aaron	28
	Patrick	30
	Esther	35
	Nora	36
	Charlie	60

24. (4) 1,000
4,000 1991 value
– 3,000 1990 value
1,000

25. (5) $\frac{1}{2}ab$ Area of a triangle = $\frac{1}{2}$ base x height
= $\frac{1}{2}ab$

Chapter 1

Looking at Word Problems

THE NUMBERS IN YOUR LIFE

$\begin{array}{r} 475 \\ -\,390 \\ \hline \end{array}$	Out of a total of 475 people who signed a petition, 390 were women. How many men signed the petition?

How are these two math problems the same? How are they different?

The big difference is that the first problem does not use any words. It simply gives you two numbers and tells you to subtract.

The second problem is a **word problem**. It describes a situation and asks a question. It gives you two numbers, just as the first problem does. However, in a word problem **you** must figure out whether to add, subtract, multiply, or divide.

For both of these problems you would subtract to get a correct answer. But with a word problem you need to take other steps **before** you subtract. Figuring out these steps is what makes word problems challenging.

Imagine driving along a highway, trying to figure out how many miles you have driven. You know the total distance to Mayfield, your destination, is 15 miles, and you see a sign that says, "Mayfield: 6 miles ahead."

The sign is giving you the information you need to find out how far you have driven. However, the sign does *not* tell you to subtract those 6 miles from the total of 15 miles. You must decide for yourself what to do with the numbers.

The number problems you come across in your everyday life are usually similar to word problems. You solve math problems with a little bit of common sense and a little bit of "number sense."

You already have acquired a lot of common sense. How do you get "number sense"? By working with numbers. By trying new things with numbers. By seeing how numbers play a part in your life.

Exercise 1

Part One

Directions: Think about the numbers you deal with each day and jot them down.

Your age: ____________________ The year you were born: ____________

Your height: ________________ A friend's height: _________________

Your weight last year: ________ Your weight this year: _____________

The time you wake up: ________ The time you go to bed: ____________

Your salary: _________________ Your rent: _________________________

Amount of money you have right now: ____________ Amount of money you spend on food each week: ________________

Cost of a loaf of bread: ________ Cost of a gallon of milk: ____________

Part Two

Directions: Use the numbers above to write five word problems. Make them as easy or as hard as you like—because you won't have to solve them! An example is given below to help you get started.

Example: Each morning I wake up at 7:00 A.M., and I go to bed at 10:00 P.M. each night. How many hours am I awake each day?

Answers will vary with each person.

READING WORD PROBLEMS

When people read word problems, they think, "Hey, this is a math problem. I'd better concentrate on the numbers." When you read a word problem for the first time, concentrate on the **story**, not the numbers. If you first picture what is "going on" in the problem, you can more easily and accurately figure out what to do with the numbers.

Look at the word problem below. Then, **without using any numbers**, answer the questions that follow.

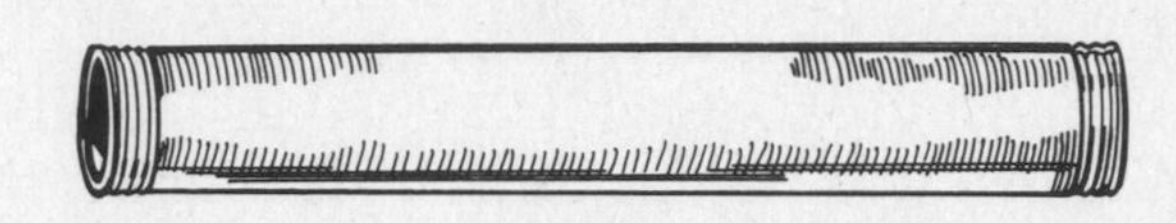

A welder needed a pipe 30 inches long to fit a newly installed sink. He cut the piece he needed from a $3\frac{1}{2}$-foot pipe. How many inches long was the remaining piece of pipe?

What did the welder need?

Where did he get it?

What is the question?

Now use your answers to reword the problem **using no numbers**. Replace the numbers with general words like *some, several, a few, a certain number of*, and *more.* You'll probably come up with something like this:

A welder needed *some* pipe. He cut it from *another* pipe. *How much pipe was left over?*

Can you picture what is happening in this problem? Before you focus on numbers and feet and inches, ask yourself, "What is going on in the problem?"

Exercise 2

Directions: Practice reading word problems without concentrating on the numbers. On a separate sheet of paper or out loud with a partner, reword each problem **without using any numbers**. Replace the actual numbers with words like *some*, *several*, *a few*, and *more.* Try to picture what is happening in each story. **Do not solve the problems.**

Example: Paul offers Naomi $.50 for every mistake she can find in his essay. After reading it carefully, Naomi finds 13 mistakes. How much money does Paul owe Naomi?

Paul offers Naomi some money for every mistake she finds. She finds many mistakes. How much does Paul owe Naomi?

1. The Crate Company hired 17 new men to work in its warehouse division. It also hired 25 women for the inventory department. How many people were hired altogether?

2. A group of 72 committee members wanted to break down into smaller work groups. If they wanted a total of 8 groups, how many committee members made up a group?

3. Over a 12-hour period, a nurse at the Alcohol Abuse Hot Line received 35 calls. If she spent an average of 7 minutes on each call, approximately how many hours in all did she spend on the telephone?

4. For the past 7 games, batter José Alvarez has had 3 hits, 5 hits, 7 hits, 0 hits, 2 hits, 1 hit, and 0 hits. What is his average number of hits for the last 7 games?

5. A tailor uses up twelve dozen spools of thread each week. If each spool costs $.35, how much does the tailor spend on thread each week?

D

6. Seventeen mail carriers each worked an 11-hour day on Tuesday, due to a holiday backlog of mail. If each mail carrier is paid $5.75 per hour, what was the total paid in wages for these seventeen carriers on Tuesday?

7. When Todd works overtime at the plant, he earns $12.50 per hour. His regular hourly wage is $9.90. How much more does Todd earn per hour when he works overtime?

8. A woman started a trip driving 55 mph, and she drove for 3 hours at this rate. Toward the end of her trip she picked up speed and drove 62 mph for an hour and a half. How many miles did the woman drive altogether on this trip?

Answers begin on page 178.

MULTIPLE-CHOICE PROBLEMS

At Chase's Market, celery is sold for $.89 per pound. A larger supermarket in the same neighborhood sells celery for $.78 per pound. If Robert bought 3 pounds of celery at the supermarket, how much money did he save by not buying it at Chase's?

(1) $.11
(2) $.27
(3) $.33
(4) $2.34
(5) $2.67

Why is it that some word problems include a choice of answers?

Do you think answer choices make the problem easier?

Word problems like the one above are **multiple choice**. (The correct answer is **(3) $.33**.) With these problems you are given a list of answers and you must choose the correct one. A list of answers does not necessarily make your job easier. You still must **solve** the word problem as you would if no answer choices were given.

DID YOU KNOW . . . ?

- Tests have a multiple-choice format because they are easier to score that way.
- On multiple-choice tests each *wrong* answer listed is carefully chosen. For example, if you added when you should have subtracted, you might find your *incorrect* answer listed as one of the choices!
- Simply guessing an answer is not a good strategy, but there are ways you can eliminate one or two of the answer choices. You'll learn more about these strategies later in this book.

For most of your work in this book you should solve the word problem without paying too much attention to the answer choices. A list of answer choices is **not** an invitation to guess blindly.

Exercise 3

Directions: First read the following word problems. Then imagine you are writing a multiple-choice test and list three possible answer choices for each problem. Make sure one of the answers is the correct one!

Example: Bill weighs 196 pounds right now. He would like to lose 16 pounds. What is Bill's desired weight?

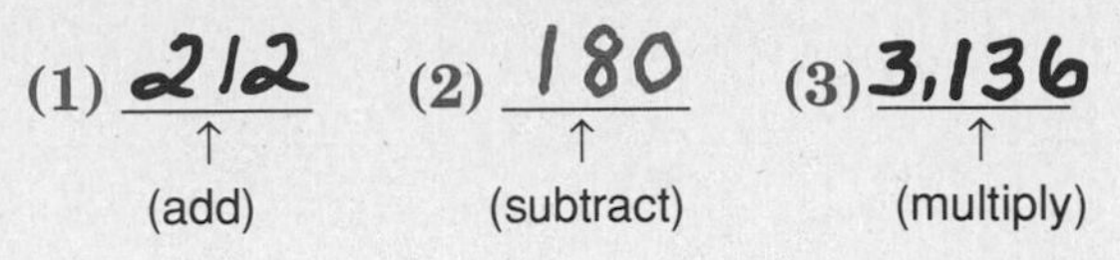

WN

1. Joanna put $127 worth of clothing on layaway with a $10 down payment. The next week she came into the store and put down another $10. Last Thursday she put down $25 more. How much more money does she owe for the clothing?

   (1) ______ (2) ______ (3) ______

2. On Halloween, Mr. Dody put together all the candy bars collected by each of his three children. Then he gave each child 1 candy bar per day for as long as the candy lasted. Patrick collected 20 bars, Peter collected 14 bars, and Jack collected 11 bars. How many days did the candy last?

   (1) ______ (2) ______ (3) ______

3. The 2 pitchers that Mrs. Sylvia is using hold $4\frac{1}{2}$ quarts and $1\frac{1}{2}$ quarts. How many more quarts does the larger pitcher hold?

   (1) ______ (2) ______ (3) ______

4. Nancy left $500 in a savings account for 1 year. The bank pays 7% interest annually. What total amount of money did Nancy have at the end of the year?

   (1) ______ (2) ______ (3) ______

5. Roberto enclosed the rectangular garden at right with wire mesh. How many feet of mesh did he need?

   (1) ______ (2) ______ (3) ______

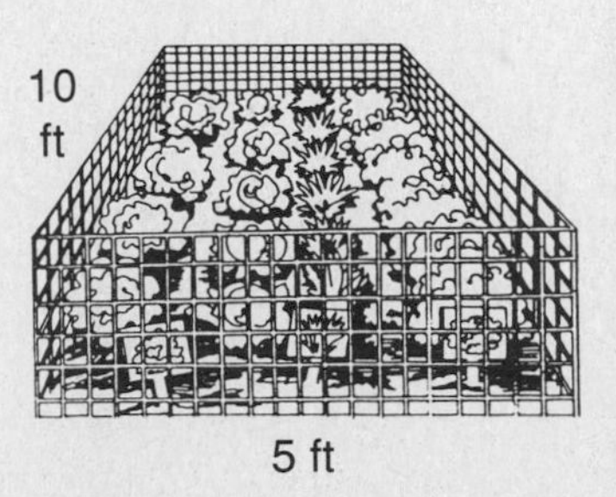

Answers begin on page 178.

MULTI-STEP WORD PROBLEMS

On Thursday, Tom collected 55 cans; on Friday, he collected 65 cans. How many cans did he collect in all?	On Thursday, Tom collected 55 cans; on Friday, he collected 65 cans. He took them to the recycling center, where he was paid $.05 per can. How much did Tom earn for the two days?

How are these two problems different?

What are the steps in solving each?

The first step in both problems is the same:

Problem 1	Problem 2
55 cans + 65 cans 120 cans	55 cans + 65 cans 120 cans

In fact, **120** is the answer to the first problem. Your only step was to add 55 and 65. However, for the second problem you need to perform another step:

Problem 2

120 cans
x .05 per can
$6.00 total

The second problem is a **multi-step** problem. This means that to get the correct answer you must perform more than one operation. Many word problems are multi-step problems. You'll get plenty of practice with them in this book.

Exercise 4

Directions: Solve the following word problems. Show your work for each step, then circle your final answer.

Example: Andrea bought 2 bunches of flowers for $2.99 each. She also bought a vase for $5.00. How much did Andrea spend in all?

Step 1: $2.99
× 2 bunches
$5.98

Step 2: $5.98 flowers
+ 5.00 vase
$10.98 in all (circled)

WN

1. During a flu epidemic at the plant, 36 of the company's 120 workers were out sick one day. To increase production the company hired 20 temporary workers for the day. How many people were working at the plant that day?

   Step 1: **Step 2:**

D

2. A customer purchases a box of toothpicks for $.72 plus a tax of $.04. She pays with a $10 bill. How much change should she receive from the cashier?

   Step 1: **Step 2:**

3. To save money Louise decides to walk instead of paying the $.75 bus fare each way to and from work. After 14 workdays, how much money will she have saved?

   Step 1: **Step 2:**

F

4. Hal works $8\frac{1}{2}$ hours per day, 5 days per week. How many hours does he work in a four-week month?

   Step 1: **Step 2:**

F

5. Sarah bought $1\frac{1}{2}$ pounds of apples and $3\frac{1}{2}$ pounds of oranges. How much did she pay?

   Step 1: **Step 2:** **Step 3:**

6. The Duffys' garden measures 20 feet by 40 feet. They use 2 scoops of fertilizer for every 100 square feet of garden. How many scoops do they need for the whole garden?

   Step 1: **Step 2:** **Step 3:**

Answers begin on page 178.

WORD PROBLEMS WITH VISUALS

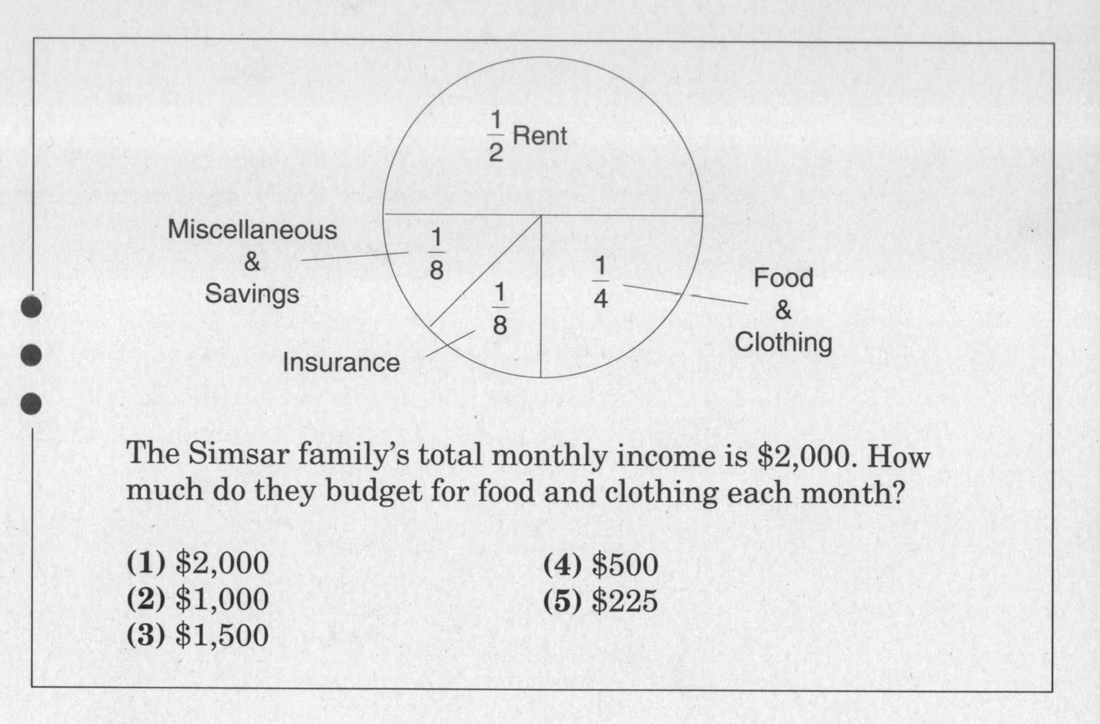

The Simsar family's total monthly income is $2,000. How much do they budget for food and clothing each month?

(1) $2,000
(2) $1,000
(3) $1,500
(4) $500
(5) $225

What information would you use to answer the word problem above?

Many word problems require you to find and use information on a graph, chart, map, or picture in order to find a solution.

To solve this problem you need to take information **from the graph** and **from the problem** itself.

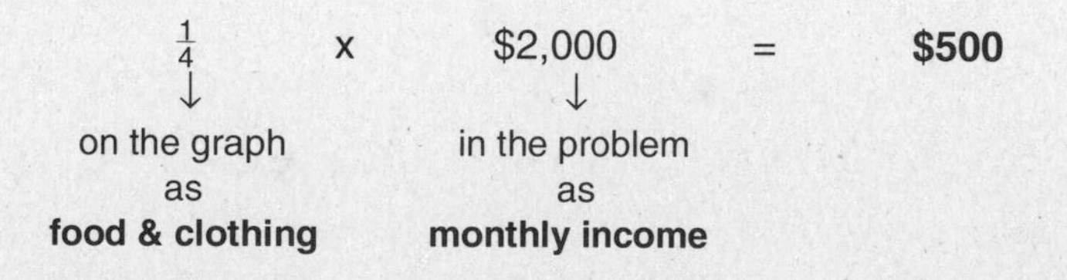

Let's look at another example of a word problem that includes a **visual**:

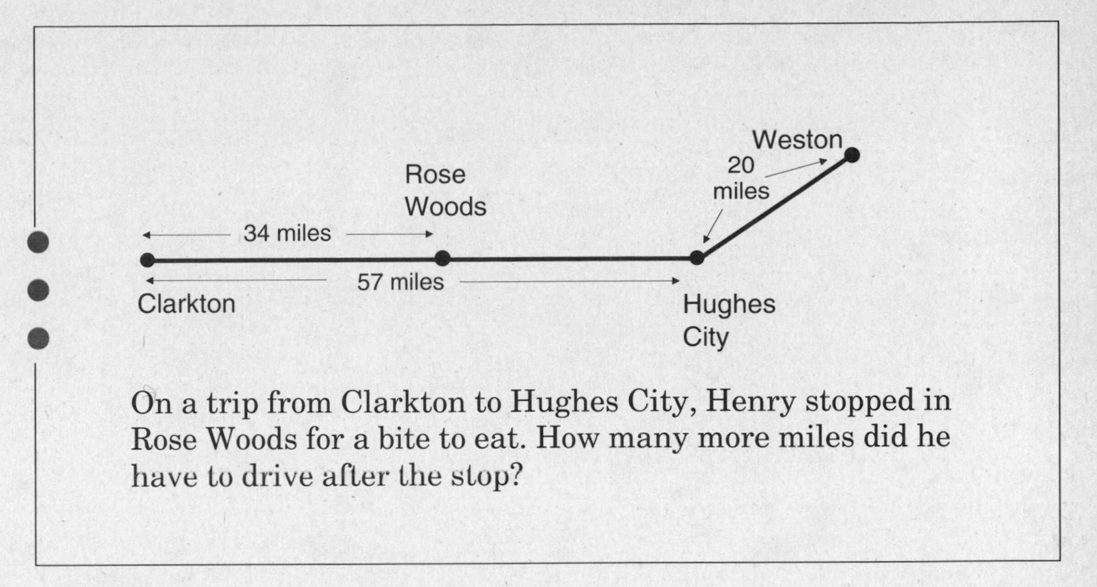

On a trip from Clarkton to Hughes City, Henry stopped in Rose Woods for a bite to eat. How many more miles did he have to drive after the stop?

As you can see, this word problem is accompanied by a map. You must look at the map and the mileage given to answer the question.

What information do you need from the map to answer the question?

What is the correct answer?

57 miles	–	34 miles	=	23 miles
↓		↓		↓
total distance Clarkton to Hughes City		Clarkton to Rose Woods		remaining distance to Hughes City

The distance from Hughes City to Weston, 20 miles, is not needed to answer the problem. Many visuals include more information than is necessary. You must be careful to choose only the numbers that you need to solve a problem.

Exercise 5

Part One

Directions: Solve the following word problems based on the chart below.

Estimated Population by Region
(in millions)

Year	North America	South America	Europe
1970	226	286	460
1980	252	365	484
1985	264	410	492
1988	272	429	497

1. How much greater was the population in Europe in 1985 than in 1980?

(1) 8
(2) 37
(3) 80
(4) 8,000,000
(5) 37,000,000

2. In 1988, how much smaller was the population of North America than the population of Europe?

(1) 225
(2) 2,250
(3) 22,500
(4) 222,500
(5) 225,000,000

3. In 1970 the population of North America was approximately what fraction of the population of Europe?

(1) $\frac{3}{4}$
(2) $\frac{1}{2}$
(3) $\frac{3}{8}$
(4) $\frac{1}{4}$
(5) $\frac{1}{8}$

4. By what percent did the population of South America grow between 1970 and 1988?

(1) 7.5%
(2) 15%
(3) 35%
(4) 50%
(5) 143%

Answers begin on page 179.

Part Two

Directions: Use the graph below to answer the questions that follow.

Quality Control Analysis
Standard Sticker Company
(first quarter 1991)

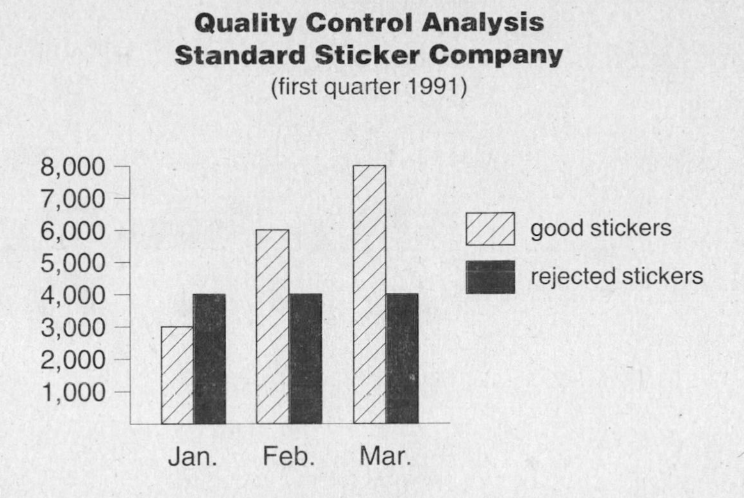

1. In January, how many more rejected stickers than good stickers were produced?

(1) 4,000
(2) 3,000
(3) 2,000
(4) 1,000
(5) 0

WN

2. How many good stickers were produced in all during the first quarter of 1991?

(1) 17,000
(2) 12,000
(3) 5,000
(4) 4,000
(5) 2,000

3. By what percent did the number of good stickers increase from January to February?

(1) 3,000%
(2) 300%
(3) 100%
(4) 50%
(5) 5%

Answers begin on page 179.

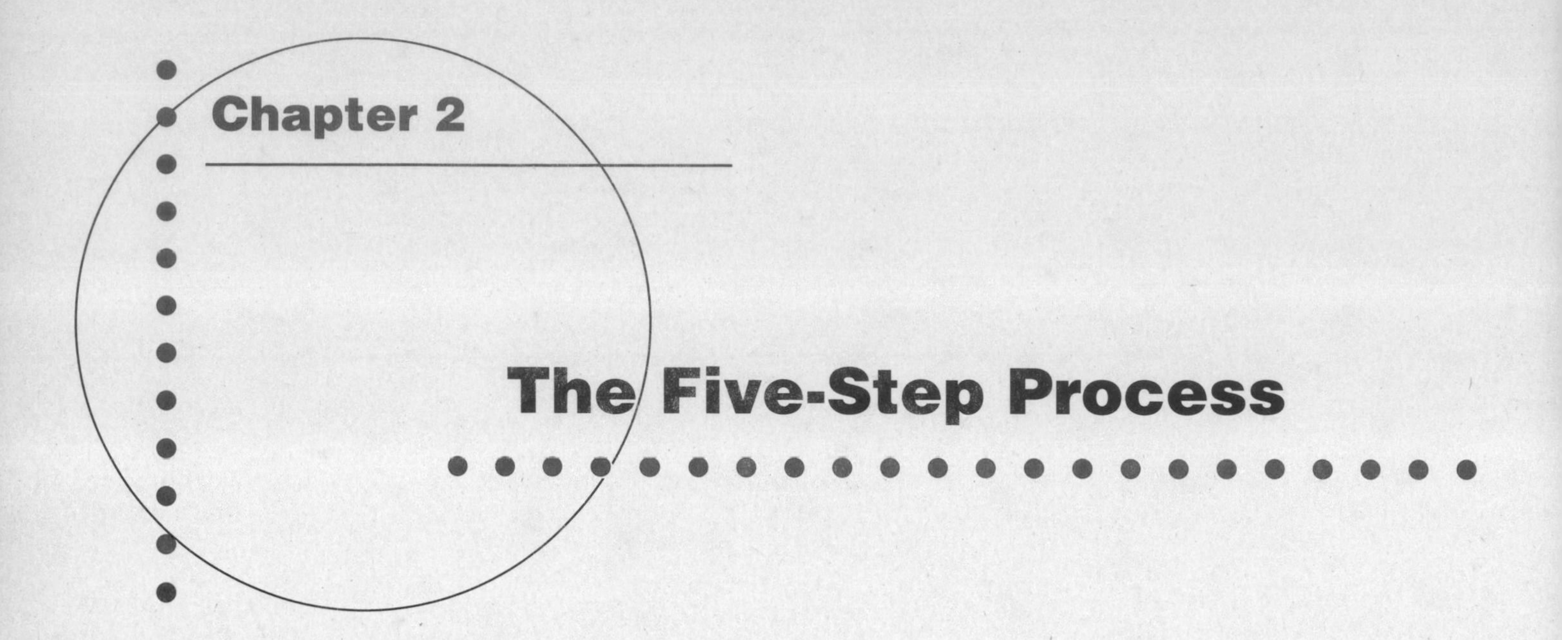

Chapter 2

The Five-Step Process

SOLVING WORD PROBLEMS

When you use a **process** to get something done, you take a series of steps, in a certain order, to reach your goal. For example, you must correctly follow a process to bake a cake, get a new job, or change the oil in your car.

WHAT WOULD HAPPEN IF . . .

- you put a cake mix in the oven *before* you added water and poured it into a pan?
- you skipped the interview and application and just showed up for a new job?
- you poured new oil into your car without emptying the old oil first?

A lot of things would probably happen if you didn't use the correct process, but you certainly wouldn't end up with a cake, a new job, or a car with fresh oil! To reach your goal you must take *every* step *in the right order*.

Solving word problems is a **process** too. Nobody can just jump in and solve a word problem in one step. Instead you must follow a series of steps, one at a time, before you can arrive at the correct solution.

Skipping steps, as you will see throughout this book, can cause serious mistakes.

For the rest of this book you will be working with different ideas to solve word problems. One of the main ideas is the five-step problem-solving process, illustrated below.

The Five-Step Problem-Solving Process

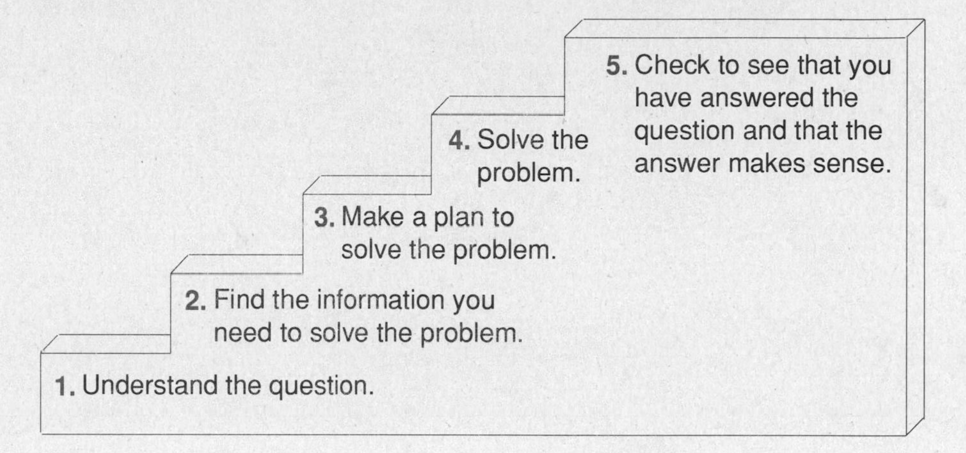

STEP 1: UNDERSTAND THE QUESTION

It took Nita's secretary 35 minutes to type a report. If the secretary can type 65 words per minute, how many words were in Nita's report?

What does the problem ask you to find?

No matter how many sentences or numbers a word problem has, it **always** asks you to find something. This problem asks you to *find the number of words in the report.*

Sometimes a problem will **ask** a question, as in the problem above. Other times a problem will **tell** you what to find.

> The bell in a church tower rings 4 times every hour from 8:00 in the morning until 8:00 at night. It then rings once per hour for the remaining 12 hours. Find how many times the bell rings in a 24-hour period.

Although there is no question mark in this problem, you *are* asked to find something. To do the problem you must *find how many times the bell rings*.

Before you can solve a problem, you must first make sure you understand the question. Otherwise you will start your work like a person who starts a trip with no idea of the destination.

Chapter 3 gives you strategies and practice to help you understand the question.

Exercise 1

Part One

Directions: Read each word problem below and write down, **in your own words**, what you are being asked to find. It may help to underline key words as you read. **Do not copy the questions and do not solve the problems.**

Example: Troy paid $.85 for each of 6 brownies at a bake sale. How much did Troy spend altogether on the brownies?

FIND: How much Troy spent on brownies

1. Ten Boy Scouts participated in a fund-raising drive one day. What amount of money did the Boy Scouts collect if each boy turned in $15?

   FIND: ______________________

2. Barbara hired 3 assistants to correct some exams. There are 63 exams to correct in all. If Barbara divides the exams evenly among the assistants, how many exams will each assistant correct?

   FIND: ______________________

3. When she worked last Saturday, Marlene earned $32.00 in tips during her 4-hour shift. If her hourly wage is $3.35, how much did she earn during that shift?

FIND: ______________________________

4. A hairdresser in Vera's Salon gave 3 haircuts at $14.00 each and 4 permanents at $20.00 each. If the salon takes 60 percent of the hairdresser's payments, how much money will it collect from the hairdresser's work?

FIND: ______________________________

Part Two

Directions: Finish writing each word problem below by adding a question. There is **no single** correct way to complete the problems, so use your imagination.

Example: Alex Moser weighs 28 pounds. His sister Juliette weighs 38 pounds. How much more does Juliette weigh than Alex?

1. Lin Chung spent $1\frac{1}{2}$ hours studying math, $\frac{1}{2}$ hour studying English, and 3 hours studying citizenship. ______________________________

2. Thirty percent of the committee members voted in favor of budget cuts. Twenty percent voted against the cuts. There are 120 members on the committee. ______________________________

3. Jen drove 130 miles in 2 hours. Sean drove 150 miles in 3 hours. ______________________________

4. A new parking lot is to be paved behind a shopping mall. The lot is 100 yards long and 36 yards wide. ______________________________

Answers begin on page 179.

STEP 2: FIND THE INFORMATION

Zena is looking at apartments to rent. One is $490 per month, including utilities. Another is $375 per month, but the tenant must pay for heat. If heating costs run about $60 per month in the second apartment, how much more can Zena expect to pay for the first apartment?

Underline any information that you need to solve this problem. Do not solve it.

Word problems do more than just ask a question. They also give you information—information that you need to solve the problem.

To solve this problem you need to know:

- the rent for the first apartment: $490
- the rent **and** the heating costs for the second apartment: $375 and $60.

Remember that the information in the problem includes **labels** as well as numbers. The numbers 490, 375, and 60 refer to *dollars*, not to inches, dogs, or miles per hour.

Chapter 4 will give you strategies and practice in finding and using information — including problems that have too much or too little information.

Exercise 2

Directions: Read each problem below and write the information that is needed to solve the problem. Don't forget to include important labels. **Do not solve the problems.**

Example: Jane drove 17 miles at 66 miles per hour. When she saw a police car, she slowed down to 55 miles per hour for the remaining 4 miles to her exit. How long did it take Jane to drive those 21 miles?

Information needed: 17 miles at 66 mph and 4 miles at 55 mph

1. Gary's job is to load canned goods into cardboard boxes. The last box he packed held cans containing 32 ounces of tomatoes, 8 ounces of pears, 16 ounces of applesauce, 16 ounces of grapefruit juice, and 32 ounces of kidney beans. What was the total number of ounces contained in this box?

Information needed: ______________________________

____

D

2. Maria took her daughter out for lunch. They bought 2 cheeseburgers, 1 small fries, 1 large fries, and 2 medium colas. Before tax, how much did the food cost?

DRIVE-IN BURGERS

HAMBURGER	$1.25
CHEESEBURGER	$1.75
FRENCH FRIES	$.75, .95
SOFT DRINKS	$.50,.75,1.00
CONES	$.75

Information needed: ______________________________

____

D

3. Robert was able to drive 211.4 miles on his last tank of gas. How many miles per gallon did he get if the tank held 12.1 gallons of gas?

Information needed: ______________________________

____

4. A candy recipe calls for $1\frac{1}{2}$ cups of sugar for every 6 ounces of chocolate. If a chef used up 36 ounces of chocolate in the candy, how much sugar did she use?

Information needed: ______________________________

____

5. A carpenter is building a planter out of railroad ties. He wants the planter to be $2\frac{1}{2}$ feet high. If he stacks up the railroad ties, and each is 4 inches high, how many ties will he need to build 1 side?

Information needed: ______________________________

____

Answers begin on page 179.

STEP 3: MAKE A PLAN

Carmelita has a sheet of dough 18 inches wide to cut into strips. Each strip must be 2 inches wide. How many strips can she get from 1 sheet of dough?

What calculations do you need to do to solve this problem?

You're right if you decided to *divide* 18 by 2. Did it take you a few minutes to decide whether to multiply or divide? If so, that's okay. You have just discovered a very important part of the problem-solving process—deciding what operation to use.

Remember, there are just four operations—addition, subtraction, multiplication, and division. However, many mistakes with word problems are made by choosing the wrong operation—for example, *multiplying* 18 by 2 in the problem above.

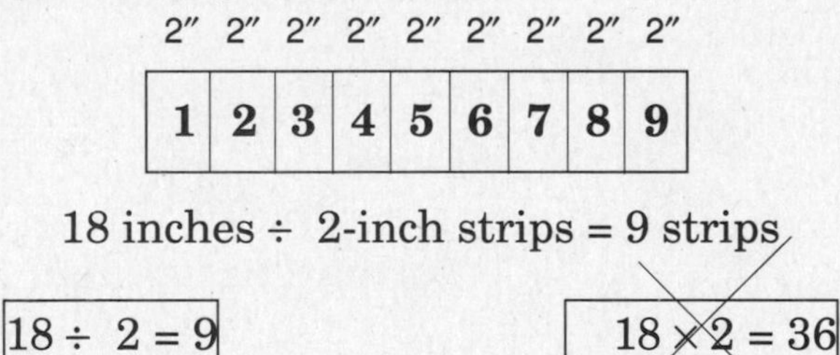

correct division | incorrect multiplication

These mistakes arise from working too quickly or not really understanding what happens to numbers when we perform different operations.

Chapter 5 will give you strategies and practice in deciding what operations to use to solve a problem.

Exercise 3

Directions: Read each problem below and decide what operation you need to perform to get the correct answer. **Do not solve the problems**.

Example: A desk that normally sells for $145 has been marked down by $30. What is the sale price of the desk?

Operation: subtract: $145 − 30

1. Roxanne weighs 120 lbs., and her daughter weighs 30 lbs. How many times heavier is Roxanne than her daughter?

   Operation: ____________________

2. In Canfield last year a total of 148 people died from handgun wounds. An additional 320 were injured by handguns. How many people in Canfield either died or were injured by handguns last year?

   Operation: ____________________

D

3. When a driver left his warehouse, he looked at his truck odometer. After he reached his destination, he read it again. How long was the trip?

   32908.4 — Before ↑ (at warehouse)

   33004.1 — After ↑ (at destination)

   Operation: ____________________

4. On a recent drug bust, law enforcement officials seized 110 kilograms of cocaine. If a kilogram equals 2.2 pounds, how many pounds were seized?

   Operation: ____________________

F

5. Saul caught a 3-pound salmon, a $3\frac{1}{2}$-pound bass, and a 2-pound bluefish on his vacation last week. How many pounds of fish in all did he catch?

   Operation: ____________________

Answers begin on page 179.

STEP 4: SOLVE THE PROBLEM

Matthew has 987,475,981$\frac{1}{8}$ yards of rope. He cuts the rope into pieces that measure 561.89341 feet each. How many pieces of rope does he have?

DON'T SOLVE THIS PROBLEM!

What seems especially hard about this problem?

The size and complexity of the numbers make this problem difficult to solve. You'll be happy to know that for this book, and for most books and tests, you will *never* have to work with numbers as large and complex as those above. In most cases the numbers you work with in your everyday life do not require lengthy calculations either.

DID YOU KNOW . . . ?

- The word problems on most mathematics tests do not require complicated and time-consuming computation.
- Most tests are actually trying to determine your problem-solving ability as well as your *basic* computation skills.
- Calculators and computers are doing more and more of the actual computation in the world today, and they help people solve math problems easily and accurately.

These facts may make you feel more at ease with math tests. However, remember that a calculator cannot choose information for you or decide what operation should be used. These "thinking skills," which are so important in word problems, are also the skills important in everyday life.

Knowing how to add, subtract, multiply, and divide accurately is *extremely* important in your work with word problems. So is working with percents, fractions, and decimals. However, this book will not concentrate on these kinds of computation, as there is not space to cover everything. To improve your math, study computation skills while you work through this book.

Chapter 6 of this book will concentrate on ways to arrive at a correct solution—and ways to avoid the traps of incorrect solutions.

Exercise 4

Directions: Practice your computation skills by doing the following problems. Remember to work neatly and carefully.

1. $3{,}987 + 452 =$

2. $452 \div 4 =$

3. $981 \times 34 =$

4. $9{,}082 - 28 =$

5. $4\frac{1}{2} + 21\frac{1}{4} =$

6. $4.56 - 3.8 =$

7. 45% of 90 =

8. $33\frac{5}{6} \times 6 =$

9. 32 is what percent of 50?

10. 66 is 30% of what number?

Answers begin on page 179.

STEP 5: CHECK YOUR ANSWER

Mr. Anton paid an 8% sales tax on his new set of utility shelves. If the price of the shelves was $60 before tax, what total amount did Mr. Anton pay for the shelves?

(1) $4.80
(2) $48.00
(3) $64.80
(4) $68.00
(5) $480

Michael, a math student, read through this problem and decided he needed to find 8 percent of $60. He did the math and came up with an answer of $4.80. Michael was working quickly, and because he saw $4.80 listed as an answer choice, he chose (1) as the correct answer.

Where did Michael go wrong?

If he had taken time to see if his answer made sense, he probably would have discovered his mistake. He gave the amount of tax, but the problem asks for the *total amount paid.*

$60	+	4.80	=	$64.80
↓		↓		↓
shelves		tax		total

To avoid mistakes like Michael's, make sure that:

- you have answered the question (*What total amount did he pay?)*
- your answer makes sense (*How could the total cost—including the tax—be less than the cost of the shelves alone?)*

Many students stop at *Step 4: Solve the Problem*. Once they have an answer, they assume it is *the right answer*. This is especially true on multiple-choice tests, where a number "on the way to the answer" may be listed among the answer choices. You can avoid errors like these if you take time to *check your answer.*

Chapter 6 will give you strategies and practice in checking your answer to see that it answers the question asked and is sensible.

Exercise 5

Directions: Each problem below has two possible answer choices. You don't need to compute to solve the problems. You should be able to see which answer choice is correct simply by deciding which choice answers the question asked and is sensible.

Circle the correct choice.

WN

1. Anthony lost 14 pounds, and he now weighs 167. What was his weight before he lost the 14 pounds?

   a) 181 pounds b) 153 pounds

2. The Myersons drove at a speed of 65 mph for 130 miles. How many hours were they driving?

   a) 65 miles b) 2 hours

WN

3. To save time, Tim had 3 friends help him deliver fliers. If each of the 4 boys delivered 72 fliers, how many fliers were delivered in all?

   a) 24 fliers b) 288 fliers

M

4. Angel added 4 cups of water to 2 cups of juice concentrate. How many pints of liquid did she have then?

   a) 6 cups b) 3 pints

5. A bus driver covered 195 miles on 1 tank of gas. The gas tank holds 15 gallons. How many miles per gallon did he get on that tank?

   a) 13 miles b) 180 gallons

F

6. Jake lost $47\frac{1}{2}$ pounds. His dad, David, lost twice that amount. How many pounds did David lose?

   a) $23\frac{3}{4}$ pounds b) 95 pounds

P

7. Sarah answered 80 percent of her math test questions correctly. The test had 15 questions on it. How many did she get right?

   a) 12 questions b) 18 questions

GE

8. How many inches of lace will Eliza need to trim a square pillow with a side that measures 6 inches?

   a) $1\frac{1}{2}$ inches b) 24 inches

Answers begin on page 180.

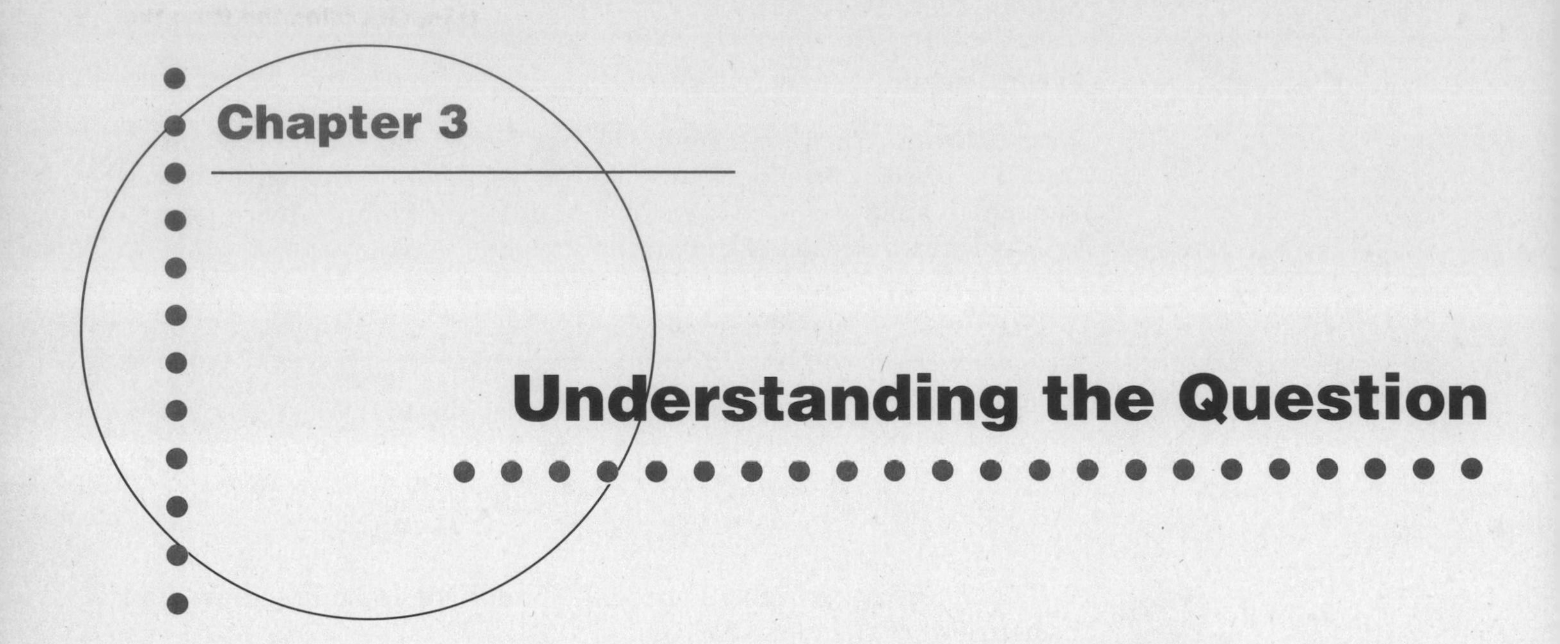

Chapter 3

Understanding the Question

READ CAREFULLY

Problem 1	Problem 2
The 3 top prizes in an essay contest were $400, $200, and $50. What was the difference between the 2 top prizes?	The 3 top prizes in an essay contest were $400, $200, and $50. What was the total given away in prizes?

Are the two problems above the same? They both refer to an essay contest, and the information provided for you is exactly the same. But the questions are not the same.

The two problems show the importance of Step 1 of the problem-solving process: understand the question. Before you can do anything with the numbers given in a problem, you must understand what you are being asked to find.

What is the answer to each problem?

What did you have to do to get each answer?

You're right if you said the answers are **$200** and **$650**.

The first problem:	**The second problem:**
$400 – 200 = $200	$400 + 200 + 50 = $650

Even though the information in the problems is the same, the answers are different because the questions are different.

Try the next exercise. It will help you practice understanding the question.

Exercise 1

Part One

Directions: Each set of problems gives the same information but asks different questions. Pay special attention to each question as you solve the problems.

1. a. On a recent trip together, Nancy drove 70 miles, Dave drove 45 miles, and Louis drove 95 miles. How many more miles did Nancy drive than Dave?

b. On a recent trip together, Nancy drove 70 miles, Dave drove 45 miles, and Louis drove 95 miles. How many miles did the 3 drive in all?

c. On a recent trip together, Nancy drove 70 miles, Dave drove 45 miles, and Louis drove 95 miles. What was the average number of miles driven by the three?

d. On a recent trip together, Nancy drove 70 miles, Dave drove 45 miles, and Louis drove 95 miles. If their total trip was going to be 250 miles, how many more miles did they still have to go?

GE **2. a.** A rectangular construction site is shown below. If a fencing company is to enclose the entire site, how many yards of fence will it need?

GE **b.** A rectangular construction site is shown below. Find the number of square yards in this site.

WN **c.** A rectangular construction site is shown below. How many times longer is the site than it is wide?

D **d.** A rectangular construction site is shown below. If each square yard was purchased for $1.25, what was the cost of the whole site?

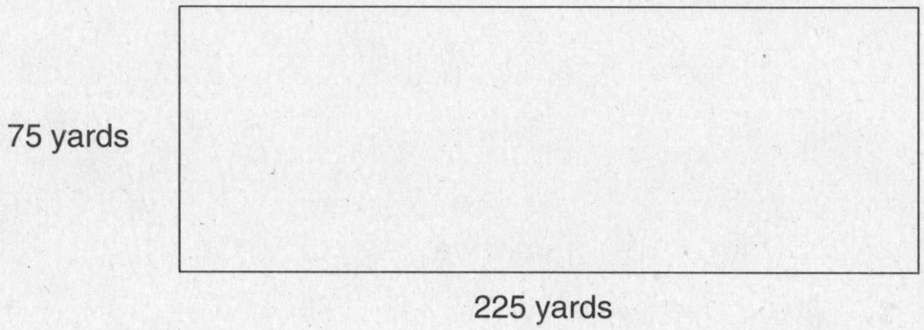

75 yards

225 yards

Part Two

Directions: You are going to help write some word problems. Read the information and then write two questions based on the information. Finally, solve the problems.

Example: Nedra worked for 13 hours over the weekend. Last weekend she worked only 5 hours.

Question 1: How many more hours did she work this weekend than last?

Solution: 13 – 5 = **8 hours**

Question 2: How many hours in all did she work over the 2 weekends?

Solution: 13 + 5 = **18 hours**

1. Both Mr. and Mrs. Phillips are employed. Mr. Phillips takes home $990 each month, while his wife takes home $1,050.

Question 1: ______________________________

____

Solution:

Question 2: ______________________________

____

Solution:

2. Three friends went out to dinner one evening. Gwen had a chef's salad for $5.50, Michael had pot roast for $7.00, and Robert had a hamburger for $4.25. They all split an ice-cream sundae for $2.00.

Question 1: ______________________________

____

Solution:

Question 2: ______________________________

____

Solution:

3. Sue and Joe Murphy spent four hours at the racetrack. During that time, Sue won $160 and Joe lost $200.

Question 1: ______________________________

____

Solution:

Question 2: ______________________________

____

(Hint: Think about each person's winnings per hour.)

Solution:

Answers begin on page 180.

SOME TRICKY QUESTIONS

Denise bought some lace to sew around 3 square pillows. Each pillow measured 4 inches on a side. How many *feet* of lace did Denise need to trim all 3 pillows?

(1) 3 **(2)** 4 **(3)** 12 **(4)** 16 **(5)** 48

Which is the correct answer to the problem above?

If you answered choice **(2)** 4, you answered correctly. However, if you answered choice **(5)** 48, you have stumbled on a common error made in word problems. **You did not answer the question that was asked.**

What steps are necessary to solve the problem?

Step 1: 4 inches × 4 sides = 16 inches

Step 2: 16 inches × 3 pillows = 48 inches

But reread the question. The question asks for the total number of *feet* of lace. Therefore, you need another step:

Step 3: 48 inches ÷ 12 = 4 feet

(12 = inches in 1 foot)

Notice how important it is to read the question **carefully**!

TIP

On some tests a word problem involving measurements may have a measurement word underlined or in *italics*. (The problem above had *feet* in italics.) Be alert for this clue that you may need to change measurement units.

To Be "Test-Wise," Know These Equivalencies

12 in = 1 ft	16 oz = 1 lb	60 sec = 1 min	2 c = 1 pt
3 ft = 1 yd	2,000 lb = 1 ton	60 min = 1 hr	2 pt = 1 qt
5,280 ft = 1 mi		24 hr = 1 day	4 qt = 1 gallon

Exercise 2

Directions: Solve the following word problems. Be sure you answer the question asked.

1. a. Celeste wants to know how long it will take her to walk from her apartment to a department store that is 4 miles away. If she walks 1 mile in 15 minutes, how many *hours* will it take her to get to the store?

(1) 1 (2) 2 (3) 11 (4) 30 (5) 60

b. Celeste wants to know how long it will take her to walk from her apartment to a department store that is 4 miles away. If she walks one mile in 15 minutes, how many *minutes* will it take Celeste to get to the store?

(1) 1 (2) 2 (3) 11 (4) 30 (5) 60

2. a. Each small truck from the Ace Removal Company can haul 500 pounds of trash at a time. On Tuesday the company has jobs to remove approximately 1,500 pounds of trash from one construction site, 500 pounds from another site, and 2,500 pounds from a third site. How many truckloads in all will Ace remove?

(1) 3 (2) 6 (3) 9 (4) 900 (5) 4,500

b. Each small truck from the Ace Removal Company can haul 500 pounds of trash at a time. On Tuesday the company has jobs to remove approximately 1,500 pounds of trash from one construction site, 500 pounds from another site, and 2,500 pounds from a third site. How many pounds in all will Ace remove?

(1) 3 (2) 6 (3) 9 (4) 900 (5) 4,500

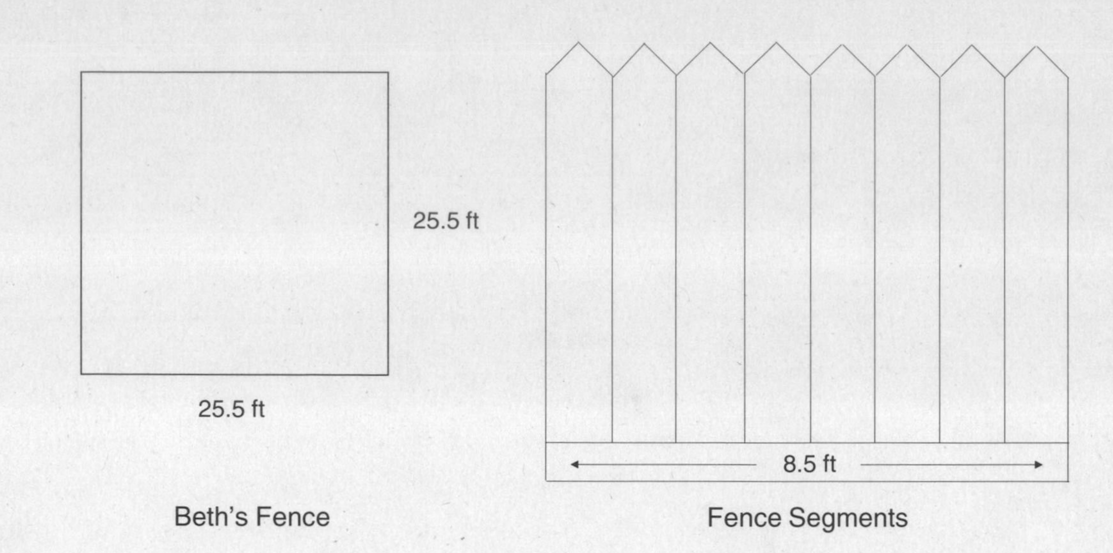

D

3. a. The fence that Beth needs to build is shown here. The fencing is sold in segments as shown and is priced at $4.70 per foot. How many segments of fence will Beth have to buy?

(1) 8.5 **(2)** 12 **(3)** 56.4 **(4)** 102 **(5)** 479.4

D

b. The fence that Beth needs to build is pictured above. The fencing is sold in segments as shown and is priced at $4.70 per foot. How much money will Beth have to spend on the fence?

(1) $8.50 **(2)** $12.00 **(3)** $56.40 **(4)** $102.00 **(5)** $479.40

F

4. a. A standard running track is 440 yards around, or $\frac{1}{4}$ mile. At a recent track meet Danita ran a 440-yard race and a 1-mile race. How many *yards* did she run in all?

(1) $1\frac{1}{4}$ **(2)** 441 **(3)** 1,440 **(4)** 1,760 **(5)** 2,200

F

b. A standard running track is 440 yards around, or $\frac{1}{4}$ mile. At a recent track meet Danita ran a 440-yard race and a 1-mile race. How many *miles* did she run in all?

(1) $1\frac{1}{4}$ **(2)** 441 **(3)** 1,440 **(4)** 1,760 **(5)** 2,200

Answers begin on page 180.

WORKING WITH "SET-UP" QUESTIONS

Maude earns \$35 per day during the week and \$60 per day on weekends. Which of the following expressions gives you the amount of money Maude earned on Monday and Saturday?

(1) $35 + 60$
(2) $60 - 35$
(3) 2×35
(4) 60×35
(5) $60 \div 35$

How does this problem look different from the other word problems you have seen?

What is the question asking you to find?

Read the problem again, **without reading the answer choices**. Does the problem seem easier now? You'll probably be relieved to learn that once you know **how** to solve the problem, you'll be able to choose the correct answer on a question like this. The problem above is a type found on many math tests. It is often called a **"set-up"** question because you have to choose one way that a problem could be "set up" for solution.

As you know, the first step in solving **any** word problem is understanding what you are being asked to find.

What are you being asked to find in this problem?

In this problem you are being asked to find an **arithmetic expression**. In mathematics an expression is a way of showing how a solution can be found. An expression is made up of two or more numbers and one or more signs of operation (+, −, ×, ÷).

Look at the problem on the top of page 33. How can you find out the amount of money Maude earned on those two days?

$$\underset{\text{Monday}}{\underset{\uparrow}{\$35}} + \underset{\text{Saturday}}{\underset{\uparrow}{\$60}} = \text{total}$$

If you look at the answer choices listed, **(1)** 35 + 60 gives you the correct expression. The other choices will not tell you how much Maude earned.

Before you move on to more challenging set-up problems, try the next exercise, which will give you more practice in working with arithmetic expressions.

Exercise 3

Directions: Choose the correct arithmetic expression for each of the following word problems. **Do not solve the problems.**

1. Carl needs an additional $240 to cover the security deposit on the apartment he wants. He has already put aside $700 for the deposit. Which of the following expressions shows the total amount of the deposit?

(1) 700 ÷ 240
(2) 700 − 240
(3) 240 + 700
(4) 240 × 700
(5) 240 ÷ 700

2. The office suite where Sondra works uses 3,000 kilowatt-hours of electricity each month. Which of the following expressions shows the number of kilowatt-hours used in a three-month period?

(1) 3,000 + 3
(2) 3,000 × 3
(3) 3,000 ÷ 3
(4) 3,000 − 3
(5) 3,000 ÷ 12

3. The temperature in Springvale yesterday was 64 degrees. Today it is 59 degrees. Which of the following expressions shows the number of degrees cooler it is today?

(1) 64 + 59
(2) 64 − 5
(3) 64 ÷ 59
(4) 59 + 5
(5) 64 − 59

4. The computer at the research center can print about 8 pages per minute. Mr. Lopez has a 70-page document to print. Which of the following expressions shows approximately the number of minutes Mr. Lopez should allow for the printing?

(1) 70×8
(2) $70 \div 8$
(3) $70 - 8$
(4) $70 \div 60$
(5) $60 \div 8$

5. A clerk at Mama's Deli wants to figure out the tax on an order of $7.10. If the food tax is 6 percent, which of the following expressions represents the tax on that order?

(1) $7.10 + .06$
(2) $7.10 - .06$
(3) $7.10 \times .06$
(4) $7.10 - .06$
(5) $.06 - 7.10$

GE

6. Which of the following expressions shows the area of Jim's garden in square feet?

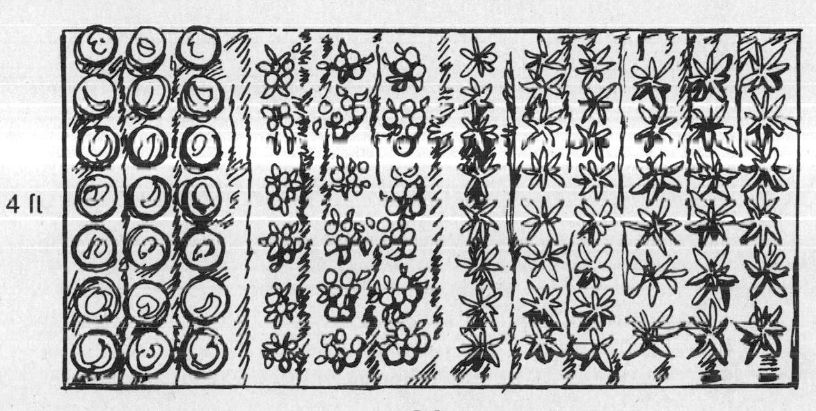

(1) $4 + 9$
(2) $4 + 4$
(3) 4×4
(4) $9 \div 4$
(5) 9×4

Answers begin on page 181.

MORE ON "SET-UP" PROBLEMS

Fifteen percent of all participants in a consumer survey said they preferred rice to potatoes. Eleven men and 25 women participated. Which of the following expressions shows the number of people in the survey who preferred rice?

(1) $15 + 11 + 25$
(2) $15\,(11 + 25)$
(3) $(.15 \times 11) + 25$
(4) $(.15 \times 25) + 11$
(5) $.15(11 + 25)$

This "set-up problem" may seem more complicated than the ones you worked with in the last exercise. As you can see, you need more than one operation to get the answer. Therefore, the answer choices listed are a little more complicated. Let's take a look at some steps in choosing the correct expression.

1. Talk about or "picture" the process needed to solve a problem. Make a statement that tells what needs to be done.	I need to add men to women and find a percent of that total.
2. Now fit the numbers given in the problem into your statement.	I need to add 11 to 25 and then take 15% of that total.
3. Write an expression using numbers and operations symbols **(before you look at the answer choice).**	.15(11 + 25) — .15 means 15%
4. Find the answer choice that matches the one you have written.	**(5)** .15(11 + 25)

Using Parentheses ( )

Were any of the operation symbols new to you? Notice that the answer choices included parentheses ( ). Parentheses mean two things in arithmetic expressions:

1. Do this first. Any computation inside parentheses should be done before any operation outside.

.15 (11 + 25)

add first

2. Multiply. When a number is written outside a pair of parentheses, you should multiply.

.15(11 + 25)
means
.15 × (36)

Using the Division Bar (—) or Slash Mark (/)

In some cases, an expression for a problem can be written in a number of ways and still have the same value.

For example, these all have the same value:
.15(11) + .15(25)
.15(25) + .15(11)
.15(25 + 11)

If you are confused by a set-up problem, think about the different ways that some expressions can be written.

Look at another example of a "set-up problem":

On Monday 291 books were checked out of the town library. On Tuesday 145 were checked out, and on Wednesday 202 books were checked out. Which of the following expressions tells the average number of books taken out over the 3 days?

(1) 291 + 145 + 202
(2) $\frac{202 + 145}{2}$
(3) $\frac{291 + 145 + 202}{3}$
(4) $\frac{291 + 145 + 202}{2}$
(5) 201 – 202 – 145

1. Figure out or "picture" the problem. • • • I need to add the numbers to be averaged, then divide by how many numbers I am averaging.

2. Fill in the numbers. • • • • • • • • • • • I need to add 291, 145, and 202, then divide by 3.

3. Write an expression. • • • • • • • • • • • $\frac{291 + 145 + 202}{3}$

4. Find the matching answer. • • • • • • • • **(3)** $\frac{291 + 145 + 202}{3}$

Did you notice how division was indicated in the problem above? A **division bar** (or fraction bar) was used:

$\frac{291 + 405 + 202}{3}$ ← this bar means divide

In some cases you may see a **slash** (/) instead of the division bar. The problem above could also be written as:

(291 + 405 + 202) / 3

Exercise 4

Part One

Directions: Write the correct expression for each of the following problems. **Do not solve the problems.**

1. Grady drove at a speed of 60 mph for the first 3 hours of a trip. Because he was late, he drove the last 2 hours at a rate of 65 mph. Write an expression that shows the number of miles Grady drove on this trip.

Expression: $(60 \times 3) + (\_\_\_\_ \times \_\_\_\_)$

2. A science class made up of 12 girls and 15 boys went on a field trip one day. The children were divided evenly into 3 groups, each group with its own adult supervisor. Write an expression that shows how many students were in each group.

Expression:______________________________

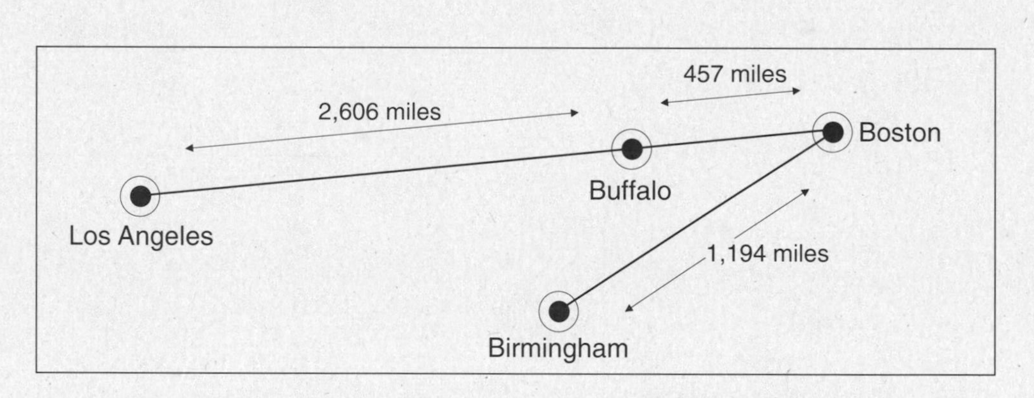

3. A salesman traveled from Birmingham to Boston, then from Boston to Buffalo in one day. The following day he flew from Buffalo to Los Angeles. Based on the map above, write an expression to show the difference in miles traveled those 2 days.

Expression:______________________________

4. For a batch of her famous spaghetti sauce Mrs. Mason bought 4 cans of plum tomatoes at $1.29 per can and 4 cans of stewed tomatoes at $.89 per can. Write an expression that shows how much Mrs. Mason paid for these ingredients.

Expression:______________________________

P

5. The Terrific Tee Shop has a sale of 35% off all T-shirts in the store. The Mugford family came in and bought 2 shirts originally priced at \$3.99. Write an expression that shows how much money the Mugfords *saved* during this sale.

   Expression:__________________________________

6. Juan and Miriam wanted to put a rope fence around the new grass they had just planted. The sides of the triangular area measured 4 feet, 9 feet, and 16 feet. Write an expression that shows how much they spent on rope if each foot of rope costs \$.11.

   Expression:__________________________________

Part Two

Directions: Choose the correct arithmetic expression for each problem. **Do not solve the problems.**

1. The members of David's bicycle club rode 13 miles on Thursday, 11 miles on Friday, 15 miles on Saturday, and 10 miles on Sunday. Which of the following expressions shows the *average* number of miles each member of the club rode on these days?

   (1) $13 + 11 + 15 + 10$
   (2) $\dfrac{13 + 11 + 15 + 10}{2}$
   (3) $\dfrac{13 + 11 + 15 + 10}{4}$
   (4) $4(13 + 11 + 15 + 10)$
   (5) $2(13 + 11 + 15 + 10)$

2. Most doctors recommend that adults consume no more than 3,300 milligrams of sodium per day. This morning Mr. Shreve ate 2 muffins, each containing 390 milligrams of sodium. If Mr. Shreve wants to stay within the recommended guidelines, which of the following expressions shows the maximum milligrams of sodium he may consume for the rest of the day?

   (1) $3{,}300 - 390$
   (2) $3{,}300 + 390$
   (3) $\dfrac{3{,}300}{2 \times 390}$
   (4) $3{,}300 + (2 \times 390)$
   (5) $3{,}300 - (2 \times 390)$

WN

3. During the months before it went out of business, Powell's Party Supplies experienced a severe decrease in sales. Based on the graph to the right, which of the following expressions shows the difference in dollar sales between July and October?

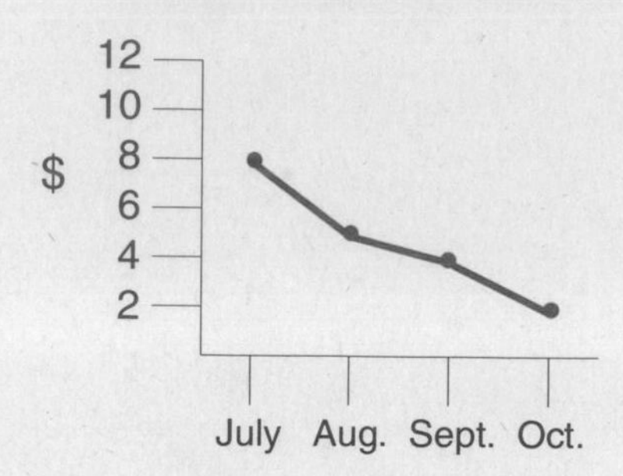

(1) 8 – 5 – 4 – 2
(2) 1,000 (8 – 5 – 4 – 2)
(3) 8 – 2
(4) 1,000 (8 + 2)
(5) 1,000 (8 – 2)

4. Maureen withdrew $50 from her savings account on Thursday afternoon. On Friday morning she deposited $110. If Maureen's account balance on Thursday morning had been $890, which of the following expressions shows her account balance on Friday afternoon?

(1) 890 – (50 + 110)
(2) (890 – 50) + 110
(3) 890 – 50 – 110
(4) 50 + 110 + 890
(5) 50 – 110 – 890

F

5. The population of India in 1988 was approximately 800,000,000 people. The population of the United States was about $\frac{1}{3}$ that number. Which of the following expressions shows the combined population of India and the United States?

(1) $\frac{1}{3}$ + 800,000,000
(2) $\frac{1}{3}$(800,000,000)
(3) $\frac{1}{3}$(800,000,000) + 800,000,000
(4) 3(800,000,000) + 800,000,000
(5) 3(800,000,000)

6. A carpet installer wants to cover $\frac{1}{3}$ of a rec room floor with industrial-strength carpet. The rest of the floor will be tile. If the floor is 10 feet wide and 24 feet long, which of the following expressions represents the number of square feet of carpet needed?

(1) $10 \times 24 \times 3$
(2) $\frac{1}{3}(10 \times 24)$
(3) $\frac{1}{3}(10 + 24)$
(4) $3(10 \times 24)$
(5) $\frac{3}{10 \times 24}$

Answers begin on page 181.

MIXED REVIEW

Directions: Choose the correct answers for the following problems.

1. A train traveled at 100 mph for 2 hours, then went an additional 2 hours at 85 mph. Which of the following expressions gives the number of miles the train traveled?

(1) $100 + 85$
(2) $4(100 + 85)$
(3) $\frac{100}{2} + \frac{85}{2}$
(4) $(2 \times 100) + (2 \times 85)$
(5) $\frac{2 \times 100}{2 \times 85}$

2. In 1987 1,043 law enforcement officers were employed in Miami, Florida. In addition, there were 357 civilian law enforcement employees. The total number of law enforcement personnel was down 34 people from the previous year. Which of the following expressions shows the number of law enforcement employees in 1986?

(1) $1,043 - 34$
(2) $(1,043 + 357) + 34$
(3) $(1,043 + 357) - 34$
(4) $1,043 + 34$
(5) $34(1,043 + 357)$

3. Sarah's time sheet is shown here. How many shifts did she work if a shift is 4 hours?

Sarah Hazelton
Employee #2213

Day	Hours
Monday	4
Tuesday	8
Wednesday	8
Thursday	12
Friday	12
Saturday	4

(1) 4
(2) 6
(3) 12
(4) 48
(5) 52

4. Rafael jogs 6 nights per week. The calendar where he records his mileage is shown here. Find the average number of miles he ran Monday through Saturday.

S	M	T	W	T	F	S
1 0	2 6.2	3 4.2	4 6.2	5 5.4	6 6.2	7 7.8
8	9	10	11	12	13	14
15	16	17	18	19	20	21
22	23	24	25	26	27	28
29	30	31				

(1) 5.4
(2) 6
(3) 9
(4) 18.4
(5) 36

5. A total of 12,000 people are expected to see the new museum exhibit. Eight thousand of these people will be coming on tour buses. What percent of the people will be from the tour buses?

(1) 8%
(2) 12%
(3) $33\frac{1}{3}$%
(4) $66\frac{2}{3}$%
(5) 100%

6. A customer at Davio's bought 2 suits priced at $220 apiece. If the clothing tax was 7%, which of the following expressions shows the total amount the customer paid for the suits?

(1) $.07 \times 220$
(2) $(2 \times 220) + .7(2 \times 220)$
(3) $(2 \times 220) + .07(2 \times 220)$
(4) $.17(2 \times 220)$
(5) $.07(2 \times 220)$

7. Mr. and Mrs. Stern sold their home for $62,500. Out of that money they had to pay the realtor a 6% commission. They took the balance of the money and put $13,000 down on a new house. How much money could they then put in the bank?

(1) $3,750
(2) $45,750
(3) $46,250
(4) $49,500
(5) $52,000

M

8. If a tailor cuts a 25-foot piece of fabric into strips that measure 4 <u>inches</u> wide, how many strips of fabric will he have?

(1) $6\frac{1}{4}$
(2) 8
(3) 75
(4) 100
(5) 1,200

M

9. Beverly poured 4 cups of flour, 2 cups of sugar, and 6 cups of milk into a bowl. How many total *pints* did she mix together?

(1) 4
(2) 5
(3) 6
(4) 12
(5) 18

GE

10. Sketches of Ellen's gardens are shown below. The insect killer she plans to use covers 10 square *yards* per jar. How many jars does Ellen need to buy?

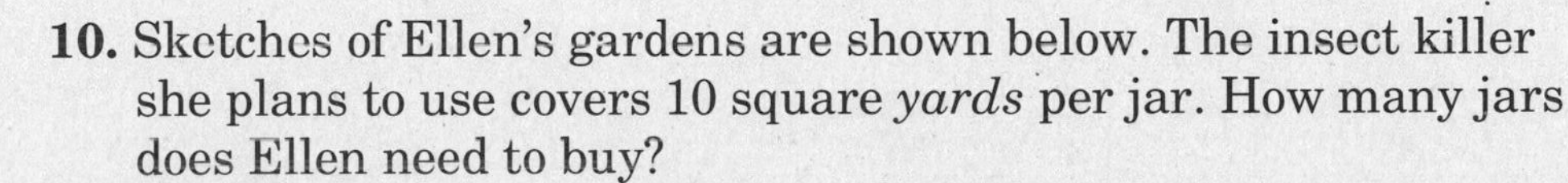

(1) 4.5
(2) 5
(3) 9
(4) 9.5
(5) 45

10 ft
15 ft

Answers begin on page 182.

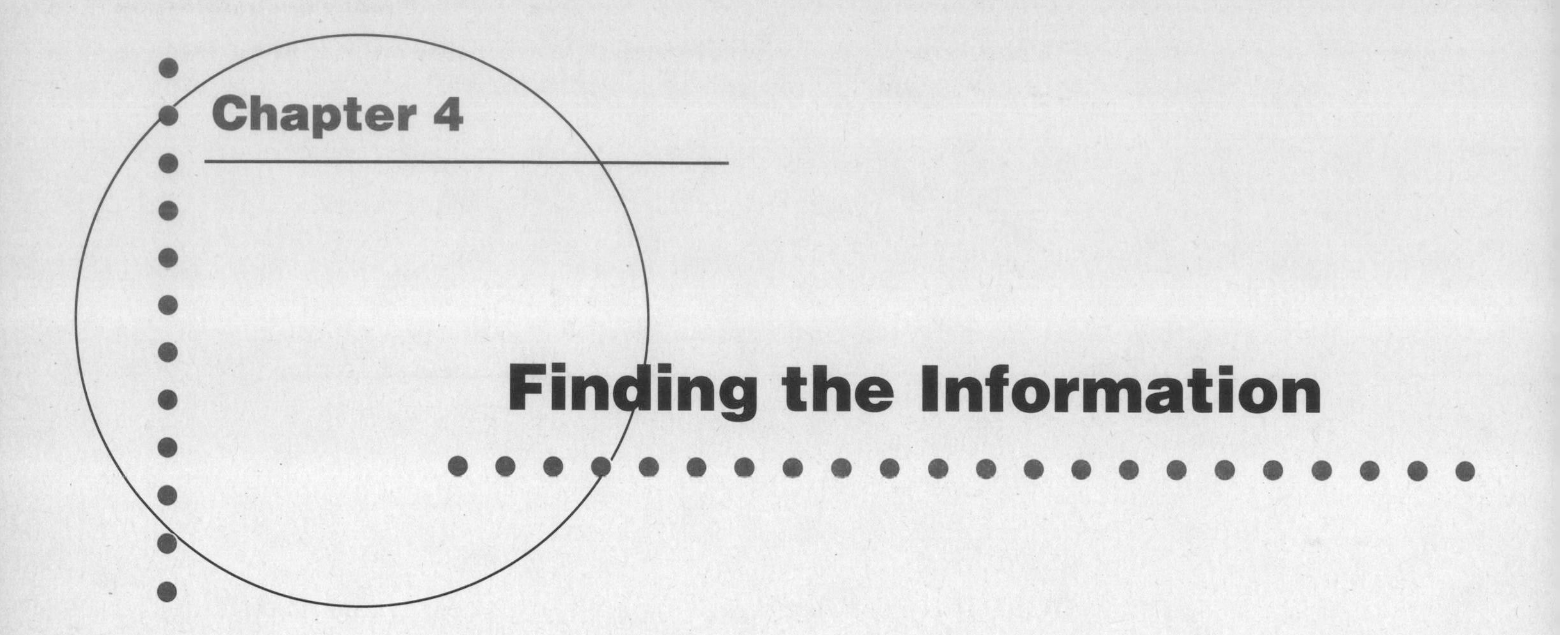

Chapter 4

Finding the Information

LOOKING AT LABELS

Mrs. Martínez purchased 4 oranges, 3 pears, 6 pounds of ground beef, and 5 red apples. How many pieces of fruit did she buy in all?

If you read this problem quickly and just add up all the numbers, you won't get the right answer. To solve this and other word problems correctly, you must pay attention to **labels**.

What are the labels in the problem above?

Which of the following is the correct solution?

Solution 1	Solution 2
4 oranges	4 oranges
3 pears	3 pears
6 pounds of beef	+ 5 apples
+ 5 apples	12
18	

The words *oranges, pears, ground beef, and apples* are important **labels** in this problem.

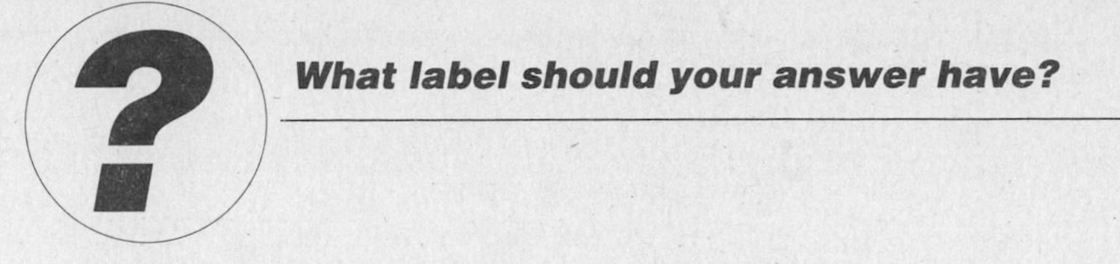

The problem asks for the number of pieces of *fruit*; therefore, you should add up the number of apples, pears, and oranges. Ground beef is not a fruit, so it should not be added in the total.

12 pieces of fruit is the correct answer.

Exercise 1

Directions: For every number in the word problems below, circle the label that belongs with it. Then solve each problem, *including a label in your answer.*

Example: Ron divided his classroom into 5 groups. Each group contained 6 students. How many students were in Ron's classroom?

5 groups x 6 students = 30 students

1. The orchard Nancy wanted to purchase contained 11 acres of apple trees. Approximately 120 trees grew on each acre. How many trees did the orchard contain?

2. On Dan Brunner's used-car lot there are 4 Chevrolets, 10 Fords, 10 Chryslers, and 9 Toyotas. How many cars are on Dan's lot altogether?

3. On Vince's recent airplane trip he covered 900 miles in $2\frac{1}{2}$ hours. On average, how many miles per hour did the airplane fly?

4. If 1 meter equals approximately 39 inches, about how many inches are in 10 meters?

5. For the room shown here, how many square feet of linoleum would you need to cover the floor?

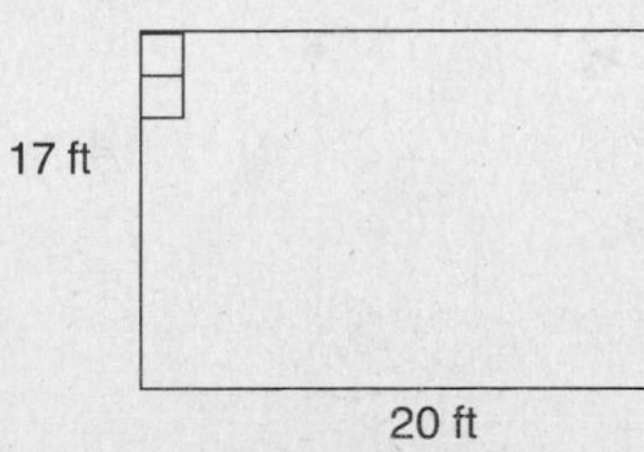

Answers begin on page 182.

FINDING "HIDDEN" INFORMATION

During a particularly rainy April, Charleston received 3.5 inches of rain in 1 week. What was the average number of inches per day?

Sometimes information in a word problem can be "hidden"; in other words, the numbers needed to solve the problem might not be obvious.

Take a minute to think "out loud" about how you might solve the problem above.

1. Perhaps you said something like this:	"I will divide the total number of inches by the number of days in a week to get the average inches per day."
2. Then you put in the numbers.	I need to divide 3.5 by 7.
3. Now, do the computation.	$3.5 \div 7 = .5$ inches

Where did the number 7 come from? It wasn't in the problem, was it?

Charleston received 3.5 inches of rain in 1 week. 1 week = 7 days

You needed to remember that there are 7 days in a week and to use that information in the problem.

Take a look at another word problem with hidden information.

A mail carrier on the morning route delivered letters to ninety houses. Her partner on the afternoon route delivered to twice that many. How many houses in all did the two mail carriers deliver to?

Is this a math problem?

Where are the numbers?

Don't give up—this problem has plenty of numbers to work with.

- Circle the number of houses the morning carrier delivered to.
- Now circle the clue that tells you how many houses the afternoon carrier delivered to.

Did you circle *ninety* and *twice that many*? This is the information that you need to solve the problem correctly.

$$90 + (2 \times 90) = 270 \text{ houses}$$

morning ↗ ↖ afternoon

Take a look at one more word problem with "hidden" information. Notice how important it is to **read carefully.**

> In all but 3 months of 1990, Sasha sold 1 dictionary per month in her job as a door-to-door salesperson. If Sasha earned \$4.50 for each dictionary sold, what were her profits on 1990 dictionary sales?

In how many months did Sasha sell a dictionary?

Multiply this number by \$4.50.

If you thought Sasha sold 3 dictionaries, you did not find the hidden information in this problem. If she sold dictionaries *all but 3 months,* for how many months did she sell dictionaries?

```
  12 months
−  3 months without sale
   9 months with sales
```

Sasha sold 1 dictionary each of the 9 months.
Since $9 \times 4.50 = 40.50$, Sasha earned **\$40.50**.

What was the hidden information in this problem?

First of all, you needed to know what the phrase *all but* meant; second, you needed to know that there are 12 months in a year.

TIP

Read all the information carefully to see if a problem has a number in word form (*eight*) or a "hidden number" (*a week* = *7 days*).

Exercise 2

Directions: Solve the following word problems. Read carefully and watch out for "hidden" information.

WN

1. Richard's monthly salary is twice what he pays for room and board at the Daze Inn. If he pays $440 per month at the inn, what is his monthly salary?

WN

2. Wanda walked 15 blocks from her home to the coffee shop, then walked the same distance back. If she walks at a rate of 15 blocks per hour, how much time did she spend walking?

3. Mary did a survey of her coworkers to see how they used the lounge at work. She found that 3 people took naps, 8 people came in to talk with coworkers, 17 people came in to smoke cigarettes, and 10 people bought coffee. How many people in all used the lounge that day?

WN

4. A cook at Franco's Café baked all but 10 pounds of the potatoes he had on hand for Saturday night dinners. He started out with 5 bags of potatoes, each weighing 8 pounds. How many pounds of potatoes did the cook bake for Saturday night?

D

5. For an art project, Francine buys 3 sheets of red felt, 10 sheets of yellow felt, and a dozen sheets of black. If each sheet of felt costs $.55, how much does Francine spend on felt?

D

6. At the Redville Cinema, one hundred eighteen people each paid $5.00 to see the early evening show. Seventy-two people paid the same amount to see the late show. How much money did the cinema take in for these two shows?

7. Sixty-four percent of all the dairy cows on U.S. farms are located in the Midwest. Based on the graph to the right, how many dairy cows were in the Midwest in 1987?

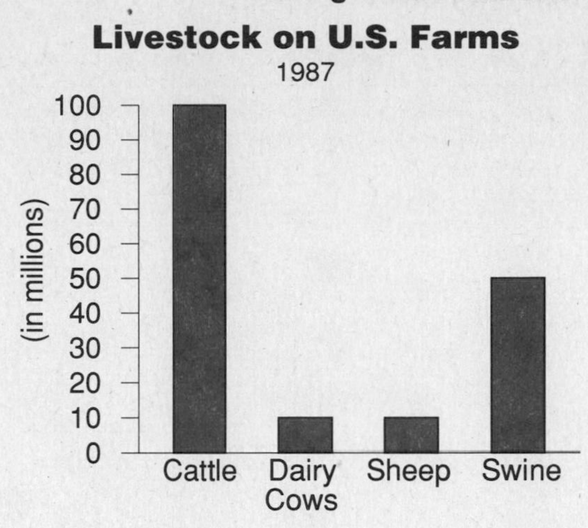

M

8. Belinda works hard to keep her commitment as a caseworker and as a good parent. She leaves for work at 8:00 in the morning and gets home at 6:00 in the evening. She finds that 6 hours of sleep per night is plenty for her. How many hours does this leave for herself and her family?

M

9. To fence his garden, Julius put a stake in the ground every three feet. Then he tied string between the stakes. If Julius's garden is ten *yards* around, how many poles did he need?

GE

10. Max is building a patio for Mrs. López. He will charge her $6 per square foot of patio, including all labor and materials. If Mrs. López wants a patio as shown, how much will Max charge her?

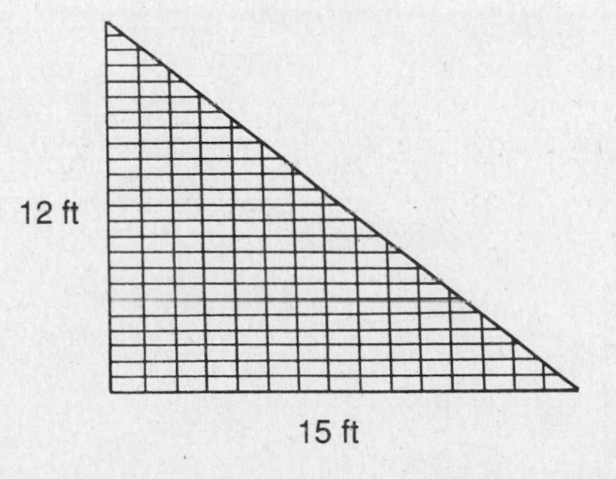

Answers begin on page 182.

EXTRA INFORMATION

A salesman was paid $3.50 per hour, plus a bonus of $2.50 for every sale he made. If the salesman worked ten hours and made 12 sales, how much did he get as a bonus?

(1) $25 **(2)** $30 **(3)** $35 **(4)** $65 **(5)** $120

For this problem, review the first two steps of the five-step process on page 15.

1. What is the question asking?	It is asking how big a bonus the salesman got.
2. What information do you need to answer the question?	You need to know that he got $2.50 per sale and that he made 12 sales.

What's tricky about this problem? You are given **more information** than you need to solve it.

Question: How much did he get as a bonus?

Information Needed: $2.50 per sale; 12 sales

Extra Information: $3.50 per hour salary; 10 hours worked

TIP

Word problems may contain extra information. The multiple-choice answers will show choices that you could make if you mistakenly used the extra information. Be careful!

What is the answer to the problem?

You're right if you said (2) **$30**. If you thought the answer was (4) $65, go back and read the question again. Remember, you want to find only the bonus—not the total pay.

Now look at another word problem and decide what information is needed to solve it.

> To get to work each day, Eva must walk 10 minutes from her house to the train, take a 25-minute train ride, walk another 5 minutes to the bus, and ride 15 minutes on the bus. How many minutes does Eva spend riding to work each day?
>
> **(1)** 15 **(2)** 30 **(3)** 35 **(4)** 40 **(5)** 55

1. What is the question asking? • • • • • • It is asking for the number of minutes spent riding.

2. What information do you need? • • • • • You need to know that she spends 25 minutes on the train and 15 on the bus.

3. What operation do you use? • • • • • • • Add 25 + 15.

4. What is the answer? • • • • • • • • • • • **(4)** 40

If you chose **(5)** 55 as the correct answer, go back and reread the question. You probably used the extra information about Eva's *walking* time. The question asks for *riding* time.

• • • • • • • •

Exercise 3

Part One

Directions: For each problem, write the question and the necessary information in your own words. Cross out the extra information. **Do not solve the problems.**

extra information →

Example: In 1987, the estimated population of Indiana was 5,531,000 people. ~~About 700,000 of those people lived in the state capital, Indianapolis.~~ If the estimated state population was 5,490,000 in 1980, how many more people lived in Indiana in 1987 than in 1980?

Question: How many more people lived in Indiana in 1987 than in 1980?

Necessary Information: 5,531,000 and 5,490,000

1. Ron is trying to decide whether or not to buy a new car from Milton's Chevrolet Dealership. The car he wants has a base price of $7,995, with an extra $700 for air-conditioning, $350 for a sunroof, and $200 for rustproofing. If Ron decides to buy the car with a sunroof and rustproofing, what will it cost him?

   Question: ____________________

   Necessary Information: ____________________

2. Officer Perry stopped a car that was traveling at a speed of 82 miles per hour. Three hours later he spotted the same car traveling at 70 mph. If the speed limit was 55 mph, by how much had the driver exceeded the limit the first time he was stopped?

   Question: ____________________

   Necessary Information: ____________________

3. An adult's ticket to the carnival is $5.50. A child's ticket is $4.00. If 100 children and twice as many adults paid for the carnival on Sunday, how much money was collected for adult tickets?

   Question: ____________________

   Necessary Information: ____________________

4. Of the 22,500 ice cream cones served by Littlefield's Dairy Bar last summer, 40% were small and 35% were large. If each large cone cost $.55, how much money did the dairy bar take in for large cones?

   Question: ____________________

   Necessary Information: ____________________

Part Two

Directions: Cross out the extra information and solve the problems.

1. The Arzadon family drove 396 miles in 7 hours. They used 12 gallons of gas. How many miles per gallon did their car get on this trip?

2. The waitresses at Flannigan's put all their tips into one community pool, then divide the tips evenly among all waitresses. One day Sandra earned $12.06 in wages and $51 in tips. Carol earned $32 in tips that day, while Madge earned $64 in tips and $12.06 in wages. How much did each waitress take home in tips?

3. How many more grams of fat does a tablespoon of mayonnaise have than a tablespoon of sour cream?

	Total Fat (grams per tbsp)	Cholesterol (milligrams per tbsp)
butter	11	31
margarine	11	0
mayonnaise	11	8
sour cream	3	5

4. Carmella enjoys watching 3 movies on television and 2 at the theater each week. Her son watches twice as many movies on TV but only half as many in the theater. How many movies does Carmella's son watch in the theater each week?

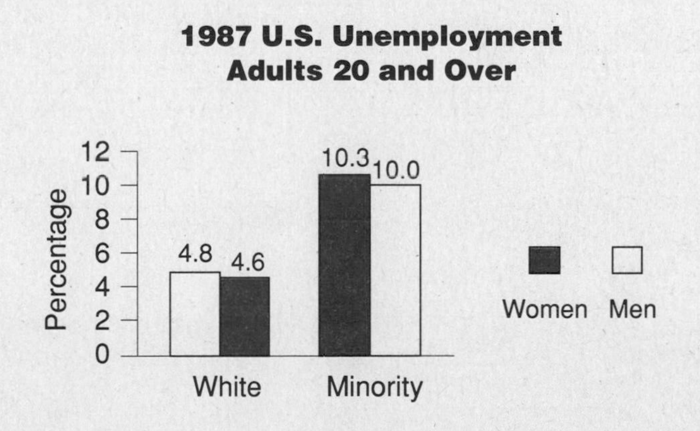

5. Based on the graph above, how much lower was the unemployment rate for minority women than minority men?

Answers begin on page 183.

NOT ENOUGH INFORMATION

At Anchor Hardware, 2-inch nails are sold by the box. A carpenter came in to purchase 4 pounds of these nails. How much did he spend if a box of nails is priced at $2?

(1) $2.00
(2) $4.00
(3) $8.00
(4) $16.00
(5) not enough information is given

Did you have trouble finding the answer?
What information is missing?

Notice that the problem tells you how much a *box* costs—but not how much a *pound* costs. You do not know how many pounds are in a box. Which answer is correct?

You're right if you said **(5) not enough information is given**. This type of word problem appears on many tests, and you'll get practice with it throughout this book. Try another one:

Nathan's salary is $20,000 per year. The company he works for withholds 15% of this salary for state tax, federal tax, social security, and health insurance. How much did Nathan pay for health insurance this year?

(1) $300
(2) $3,000
(3) $6,000
(4) $9,000
(5) not enough information is given

What information is missing from the problem?

Although you are told the total percentage that is withheld from Nathan's salary, you do not know what part of that goes toward health insurance. If you chose answer **(2)**, you made a common error—you found 15% of $20,000, but you did not answer the question "How much did Nathan pay for *health insurance*?"

Take one more look at a challenging word problem. Don't be fooled by answer choice **(5)**!

A total of 25 cheese labels are printed every 3 minutes on Line A at a factory. Line B prints 740 labels every hour. When both lines run nonstop, how many more labels are printed on Line B than Line A in 1 hour?

(1) 240
(2) 500
(3) 665
(4) 715
(5) not enough information is given

TIP

Don't choose *not enough information is given* as an answer just because you are confused by a problem. Be sure you can identify *what information is missing* before you choose that answer.

At first you might be tempted to choose **(5)** as the answer. But if you work carefully, you'll see that you *are* given enough information. Use the five-step process:

1. What is the question asking?	How many more labels are printed on B than A in one hour?
2. What information do you need?	Line B = 740 labels Line A = 25 labels × 20 3-minute periods in 1 hour = 500 labels
3. What's the plan?	To find how many more, you should subtract.
4. Do the math.	740 – 500 = 240 labels
5. Check your answer.	**240** is a sensible answer because it is the only choice less than Line A. **(1)** is the correct choice.

At first you may not have realized that you *can* figure out how many labels are printed in one hour on Line A. Although this information is not stated directly, you can use the numbers provided to find the right answer.

Exercise 4

Part One

Directions: None of the following word problems can be solved because some information is missing. Read each problem and write down what information would be necessary to solve the problem.

Example: A bus traveled 55 miles per hour for a period of 4 hours. It then picked up speed to 65 mph and continued the trip. How far did the bus travel in all?

Missing Information: number of hours at 65mph

WN

1. Steven's exercise class is full on Tuesdays and Thursdays. On Wednesdays, he gets half as many participants, and on Mondays and Fridays, 25 people are exercising. How many students in all does he teach each week?

   Missing Information: ______________________________

WN

2. The monthly rent for a studio apartment in Mr. Santos's building is \$395, plus utilities. A similar apartment in a nearby building rents for \$440, including utilities. How much money would you save per month if you rented Mr. Santos's apartment?

   Missing Information: ______________________________

P

3. Sandy bought a secondhand drum set for \$179.40, including tax. She later bought some sheet music priced at \$4.00. How much did she pay altogether?

   Missing Information: ______________________________

M

4. Mrs. Sobala leaves home at 7:30 A.M. and takes a 25-minute train ride to work. On her way home she takes a bus instead, and that trip takes 35 minutes. What time does Mrs. Sobala get home?

   Missing Information: ______________________________

GR

5. Of all the residents in Lane County, how many of them voted for the tax increase?

   Missing Information: ______________________________

   ____

   ____

Lane County Residents' Response to Tax Increase

$\frac{2}{5}$ In Favor

$\frac{2}{5}$ Opposed

$\frac{1}{5}$ Undecided

Part Two

Directions: Select the right answer for each of the following word problems. Some of the problems have enough information for you to solve them, but others do not.

1. Harold started the day with \$45 in his wallet. If he bought 2 gallons of milk at \$2.00 each, then bought 14 gallons of gas, how much money did he have left in his wallet?

   (1) \$26.57
   (2) \$28.66
   (3) \$40.82
   (4) \$42.91
   (5) not enough information is given

2. A pizza parlor spends approximately $2.70 on the ingredients to make 1 large pizza. In addition, it pays about $1.10 for the labor and all other costs in making 1 large pizza. If the parlor charges $5.75 for a large pizza, what is its profit?

(1) $1.95
(2) $3.05
(3) $3.80
(4) $9.55
(5) not enough information is given

P

3. The Guerrero family's lunch bill came to $36.50 before 6 percent tax was added to the bill. If Mr. Guerrero paid for the lunch with a $50 bill, how much change did he receive?

(1) $3.00
(2) $11.31
(3) $13.50
(4) $26.00
(5) not enough information is given

M

4. A bus driver left Station A at 4:45 P.M. and drove for an hour and 15 minutes at 60 miles per hour. He stopped at Station B to refuel and pick up more passengers. He then drove for another hour at 55 miles per hour to get to Station C. What time did the bus arrive at Station C?

(1) 6:30 P.M.
(2) 7:00 P.M.
(3) 8:00 P.M.
(4) 8:30 P.M.
(5) not enough information is given

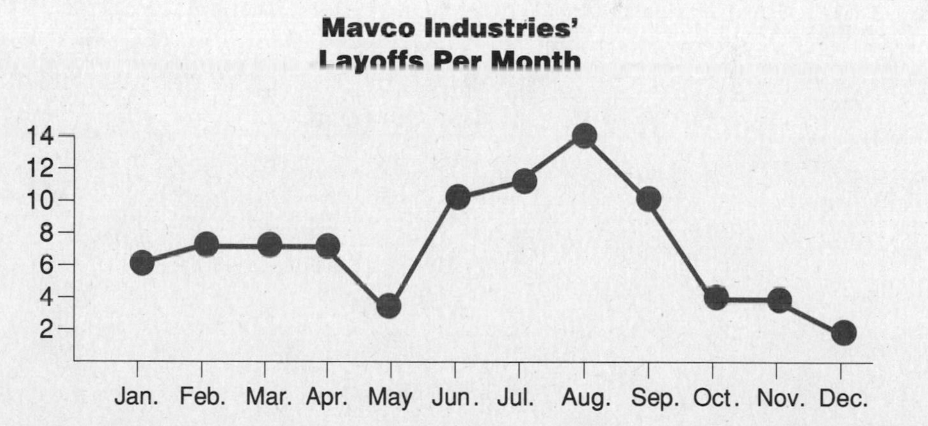

GR

5. According to the graph, how many people were employed by Mavco Industries in June, July, and August?

(1) 35
(2) 14
(3) 12
(4) 10
(5) not enough information is given

Answers begin on page 183.

WHAT MORE DO YOU NEED TO KNOW?

Nancy bought some pants at one store and a winter coat at another. Both were 20 percent off the original price. Before tax, the sale price of the coat was $140. What more do you need to know to find the total amount that Nancy spent on the coat?

(1) the sale price of the pants
(2) the original price of the pants
(3) the tax on the coat
(4) the total paid for pants and coat
(5) the original price of the coat

Is there enough information given to find how much Nancy spent on the coat?

What more do you need to know to solve the problem?

Here's another example of a word problem in which there is not enough information to get an answer. This type of problem asks you to identify what piece of additional information is needed.

If the price of the coat is $140 plus tax, then you need to know the amount of tax to find the total.

```
   $140
+    ?? ←tax
  $??? ←total
```

The answer to this problem is **(3) the tax on the coat**.

Try the next exercise to get some practice in working with this type of word problem.

Exercise 5

Directions: Solve the following problems.

1. Each crate of books that Randall fills in his packing job weighs 76 pounds. He packs a total of 3,800 pounds in 1 day. What more do you need to know to find out how many books he packs in a day?

(1) the number of pounds he packs in an hour
(2) the weight he packs in 1 hour
(3) the number of books in a crate
(4) the number of hours he works in a day
(5) the number of books he packs in an hour

2. Regina paid $2.50 for a hairbrush, $1.76 for a package of combs on sale for 10 percent off, $.69 for each of 2 barrettes, and $.95 in tax. What more do you need to know to find how much change she received from the cashier?

(1) the total cost of the barrettes
(2) the tax rate in her state
(3) the original price of the combs
(4) the amount of money she gave the cashier
(5) the total number of items she bought

D

3. After driving 70 miles, a driver has about 12 gallons of gas left in his tank. He fills it with gas priced at $1.31 per gallon. What more do you need to know to find how much he paid in all for the gas?

(1) the amount of gas his tank holds
(2) the miles per gallon his car gets
(3) the miles per hour he drove
(4) the price of his last fill-up
(5) the number of hours he drove

P

4. Twenty-three percent of the people who responded to a questionnaire said that they were pleased with the performance of the government. A total of 35,000 questionnaires were sent out. What more do you need to know to find out the number of people who said they were pleased with the government's performance?

(1) the number of people in the country
(2) the percent that said they were displeased
(3) the percent of people who responded to the questionnaire
(4) the number of answer choices for the question
(5) the percent of questionnaires sent out

5. The rug shown here is placed lengthwise on a floor that is 15 feet long. What more do you need to know to find the fraction of the floor that the rug covers?

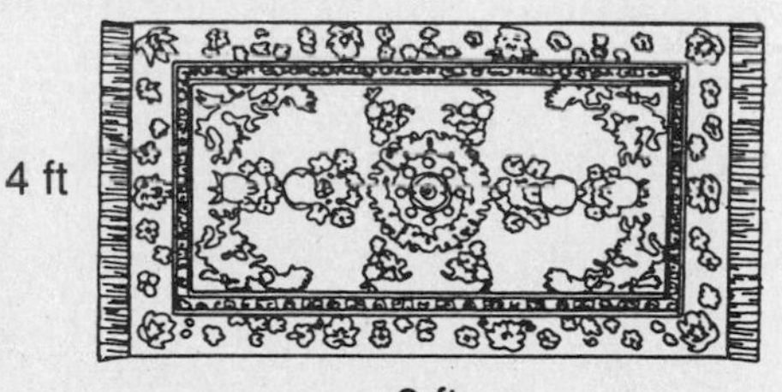

(1) the area of the rug
(2) the ratio of the length of the rug to its width
(3) the area of the floor covered by furniture
(4) the difference in length between the rug and the floor
(5) the width of the floor

Answers begin on page 184.

WORKING WITH "ITEM SETS"

Read the information and questions below. **Don't answer them yet.**

> Gwen and Richard Harris have a daughter named Maggie for whom they need to provide day care. They work an average of 20 days a month.
>
> At Little Angels Day-Care Center, the monthly charge is $320. Little Angels also charges a one-time $25 registration fee, a one-time yearly $15 insurance charge, and $5 monthly to cover the cost of field trips.
>
> However, Gwen and Richard are also considering having their friend, Roberta Mitchell, baby-sit in her home for $2 an hour.

1. How much would Gwen and Richard pay for the first month at Little Angels?

2. If Maggie stays at Mrs. Mitchell's for 10 hours a day, what would be the Harris's average monthly child-care cost?

3. After the first month, what would be the cost difference between Mrs. Mitchell's home day care and Little Angels Day-Care Center?

This type of word problem is called an **item set**. It is made up of a short paragraph followed by two or more questions.

What seems difficult about item sets?
What's the best way to handle them?

Item sets may seem hard because so much information is given. There is a lot of reading to do. In addition, having more than one question per problem may seem confusing.

Don't worry. You can successfully handle item sets by using the same skills you use when you sort necessary and unnecessary information. Here are some tips that can help:

1. **READ ALL THE INFORMATION SLOWLY AND CAREFULLY.** Don't rush into the problems. Slowly read the information once or twice until you get a "picture" of what is being presented.

2. **ANSWER EACH QUESTION SEPARATELY.** Don't try to read all the questions at once. Even though they are listed together as a set, the questions are exactly like those of regular word problems if you deal with them one at a time. Look at Question 1 from page 60 again.

How much would Gwen and Richard pay for the first month at Little Angels?

$320 + $25 + $15 + $5 = $365

↑ monthly charge ↑ registration ↑ insurance ↑ field trips

3. **AFTER READING A QUESTION, SKIM THROUGH THE PARAGRAPH AND JOT DOWN THE INFORMATION NECESSARY FOR JUST THAT QUESTION.** Use Question 2 from page 60 to see how you need to skim for the necessary information.

If Maggie stays at Mrs. Mitchell's for 10 hours a day, what would be the Harris's average monthly child-care cost?

Step 1: Find total hours

20 days × 10 hours = 200 hours

↑ Harris's work schedule ↑ (information within Question 2)

Step 2: Find total costs

200 hours × $2 = $400

4. **BE ESPECIALLY CAREFUL WITH MULTI-STEP PROBLEMS.** Some problems take two or more steps to solve. Question 3 from page 60 illustrates this.

After the first month, what would be the cost difference between Mrs. Mitchell's home day care and Little Angels Day-Care Center?

Step 1: Little Angels

$320 + $5 = $325

↑ monthly charge ↑ field trips

Step 2: Mrs. Mitchell

20 days × 10 hours × $2 = $400

Step 3: Compare

$400 – $325 = $75

Exercise 6

Directions: Read each item set carefully. For each question, jot down the information needed to solve **only that problem;** then solve it.

ITEM SET A:

At the music box factory, each assembly line manufactures 80 boxes per hour when it works at top speed. On Tuesday, eight lines worked the day shift and four lines worked the night shift. Each worker on the day shift earns $6 per hour, while the workers on the night shift earn $9 per hour. There are 4 workers on each assembly line at all times.

1. How many people worked on the assembly line on Tuesday?

Necessary Information: ______________________________

Answer: ______________________________

2. How much did the factory pay in hourly wages on Tuesday if each shift is 8 hours long?

Necessary Information: ______________________________

Answer: ______________________________

3. How many boxes are manufactured in a day shift of 8 hours on Tuesday if the lines are working at top speed?

Necessary Information: ______________________________

Answer: ______________________________

ITEM SET B:

Myra and Henry did some yard work behind their home. First they dug up a concrete area that measured 20 feet by 12 feet. They planted grass in three-fourths of that space, and they plan to put a brick patio in the rest. They will need four bricks per square foot for the patio. If they build the patio themselves, their only cost will be the bricks at $.20 each and $20 for a bag of mortar. Jack's Landscape Co. wants $320 for the entire job—including labor and materials.

20 ft

12 ft

1. How many square feet are in the grassy area of the yard?

Necessary Information: ____________________

Answer: ____________________

2. How many bricks will Myra and Henry need to build their patio?

Necessary Information: ____________________

Answer: ____________________

3. If Myra and Henry decide to do the patio themselves, how much less will it cost them than if Jack's Landscape Co. does the job?

Necessary Information: ____________________

Answer: ____________________

Answers begin on page 184.

MAKING CHARTS TO SOLVE ITEM SETS

Gloria and Robert Shaw want to buy a new car that has a list price of $9,250. Gloria's monthly take-home pay is $900, and Robert's is $1,200. They would have to make a down payment of $925 (10%) and monthly payments of $255 for 36 months.

What makes item sets difficult?

What can simplify them for you?

Many times there is so much information in an item set that it is easy to get confused. All of the facts and numbers can make it hard to select the information you need to solve each question. Here's a strategy that can help. Look at how you can organize the information from the item set into a chart. Then you can use the chart to select the information you need.

Income	Car Costs
Gloria = $900 per month Robert = $1,200 per month	• $9,250 list price • $925 down (10 percent) • 3 years of payments at $255 per month

1. How much do Gloria and Robert have to pay off after they make the down payment?

Information: $9,250; $925 down

Answer: $9,250 – $925 = **$8,325**

2. What total price will the Shaws pay for the car?

Information: $925 down; 3 years at $255 per month

Step 1: 36 months (3 years) × $255 = $9,180

Step 2: $925 down + $9,180 = $10,105

Answer: $10,105

3. How much will the Shaws have left per month after they make the car payment?

Information: $900; $1,200; $255

Step 1: $900 + 1,200 = $2,100

Step 2: $2,100 – 255 = $1,845

Answer: $1,845 per month

Exercise 7

Part One

Directions: Read the item set below and fill in the chart provided. Then answer the questions that follow.

Sasha was comparing 2 projects she had worked on. The first project took her 100 hours to complete. For this project Sasha required 10 hours of help from a computer operator who earns $7 per hour. Expenses for materials came to $124. The second project that Sasha worked on required no outside help, but it took her 20 entire workdays to complete. The materials cost the company $280. Sasha earns $9 per hour, and her workday consists of 9 hours minus 1 hour for lunch.

	First Project	Second Project
Sasha's pay at $9 per hour		
Computer operator's pay at $7 per hour		
Cost of materials		

1. For materials, how much more did the company have to pay for the second project than the first?

(1) $404
(2) $260
(3) $156
(4) $150
(5) not enough information is given

WN

2. What total dollar amount did the first project cost the company?

(1) $9
(2) $124
(3) $700
(4) $900
(5) $1,094

WN

3. What total dollar amount did the second project cost the company?

(1) $7
(2) $1,440
(3) $1,720
(4) $1,724
(5) not enough information is given

Part Two

Directions: For the item set below, fill in a chart to organize the information. Then answer the questions.

Giant Grocery sells carrots at $.69 per pound or $3.00 for a 5-pound bag. The owner of the store gives a 15 percent discount on everything *except* vegetables to local restaurants. An Italian restaurant down the street uses about 20 pounds of carrots and 25 pounds of flour each week.

cost of carrots	
restaurant purchases	
discount rates	

1. If the Italian restaurant bought bags of carrots to last a week, how much money would it save if it bought carrots in 5-pound bags?

(1) $1.20
(2) $1.80
(3) $2.31
(4) $1.71
(5) not enough information is given

2. If a nearby restaurant purchased 3 bags of sugar at $2.60 per bag, how much did it pay?

(1) $.39
(2) $1.17
(3) $6.63
(4) $7.80
(5) not enough information is given

3. How much does the Italian restaurant pay for the flour it uses each week?

(1) $3.75
(2) $7.50
(3) $8.25
(4) $9.88
(5) not enough information is given

Answers begin on page 184.

Part Three

Directions: Read the item set below and fill in the chart provided. Then answer the questions that follow.

Sue Maynard bought $5\frac{1}{2}$ yards of cotton fabric to make three baby outfits. One outfit called for 2 yards; the other two called for $\frac{3}{4}$ yard each. The fabric cost $4.00 per yard plus 5% tax.

yards needed	
cost per yard	
tax	

1. How many yards of fabric are needed for the three outfits?

(1) 9
(2) $5\frac{1}{2}$
(3) $3\frac{1}{2}$
(4) 2
(5) $1\frac{1}{2}$

2. How many yards of fabric did Sue have left after she made the 3 outfits?

(1) 9
(2) $5\frac{1}{2}$
(3) $3\frac{1}{2}$
(4) 2
(5) 0

3. How much did Sue spend on the fabric she bought?

(1) $.70
(2) $1.10
(3) $14.70
(4) $23.10
(5) $23.80

Answers begin on page 185.

MIXED REVIEW

Directions: Solve the following word problems.

WN

1. The Hair Etc. chain of hair salons gave 340 haircuts and 210 permanents in July. It gave 420 haircuts and 185 permanents in August. How many more haircuts were given in August than July?

(1) 25
(2) 80
(3) 130
(4) 365
(5) 410

2. To settle an argument among his 3 children, Bob took the 12 marbles Anna had, the 27 marbles Bea had, and the 33 marbles Gary had and divided them evenly among the children. How many marbles did each child end up with?

(1) 216
(2) 72
(3) 27
(4) 24
(5) 8

3. Before a rat was injected with Drug X, it traveled through a maze in 12.1 seconds. After the injection the same rat traveled the maze in 15.3 seconds. A second rat, with no injection, completed the maze in 15.2 seconds. How many seconds slower was the first rat after the injection than before?

(1) 3.2
(2) 3.1
(3) 3
(4) .32
(5) .31

4.

	Adult	**Child**
Friday	870	435
Saturday	1,115	224
Sunday	1,002	331

The chart shows the attendance numbers for the basketball playoffs at Milton High. Adult tickets sold for \$3.50; children's tickets sold for half that price. How much money was collected from children's tickets for all 3 days?

(1) \$5,227.25
(2) \$3,465
(3) \$1,732.50
(4) \$175
(5) not enough information is given

F

5. One week Al's Advertising delivered $\frac{1}{4}$ of its fliers on Monday, $\frac{1}{5}$ of the fliers on Friday, $\frac{2}{5}$ of the fliers on Saturday, and the rest on Sunday. What fraction of its fliers did Al's deliver on the weekend?

(1) $\frac{1}{20}$
(2) $\frac{3}{20}$
(3) $\frac{1}{5}$
(4) $\frac{2}{5}$
(5) $\frac{11}{20}$

M

6. On 1 workday Jill had 3 different meetings. One lasted only 15 minutes, another was an hour and 15 minutes long, and the last one was 55 minutes long. Which of the following expressions shows the number of *hours* Jill spent in meetings?

(1) $15 + 1.15 + 55$
(2) $\dfrac{15 + 1.15 + 55}{60}$
(3) $(15 + 75 + 55) \times 60$
(4) $\dfrac{15 + 75 + 55}{60}$
(5) $\dfrac{15 + 15 + 55}{60}$

Questions 7–8 refer to the following passage.

Ed's television is not working, and he is deciding whether to have it repaired or to buy a new set. He can have it fixed for $150. A new TV at Bert's Discount will cost him $305, plus an extra $75 for a warranty. The same set at Tech TV will cost him $400, including the warranty.

7. How much more will the TV plus warranty cost at Tech TV than at Bert's?

(1) $20
(2) $80
(3) $95
(4) $155
(5) not enough information is given

8. Which of the following expressions shows the amount of money Ed will save by getting his TV repaired instead of buying one at Bert's without a warranty?

(1) $305 – $150 – $75
(2) $305 + $75 – $150
(3) $305 – $150
(4) $400 – $150
(5) $305 + $150

Answers begin on page 185.

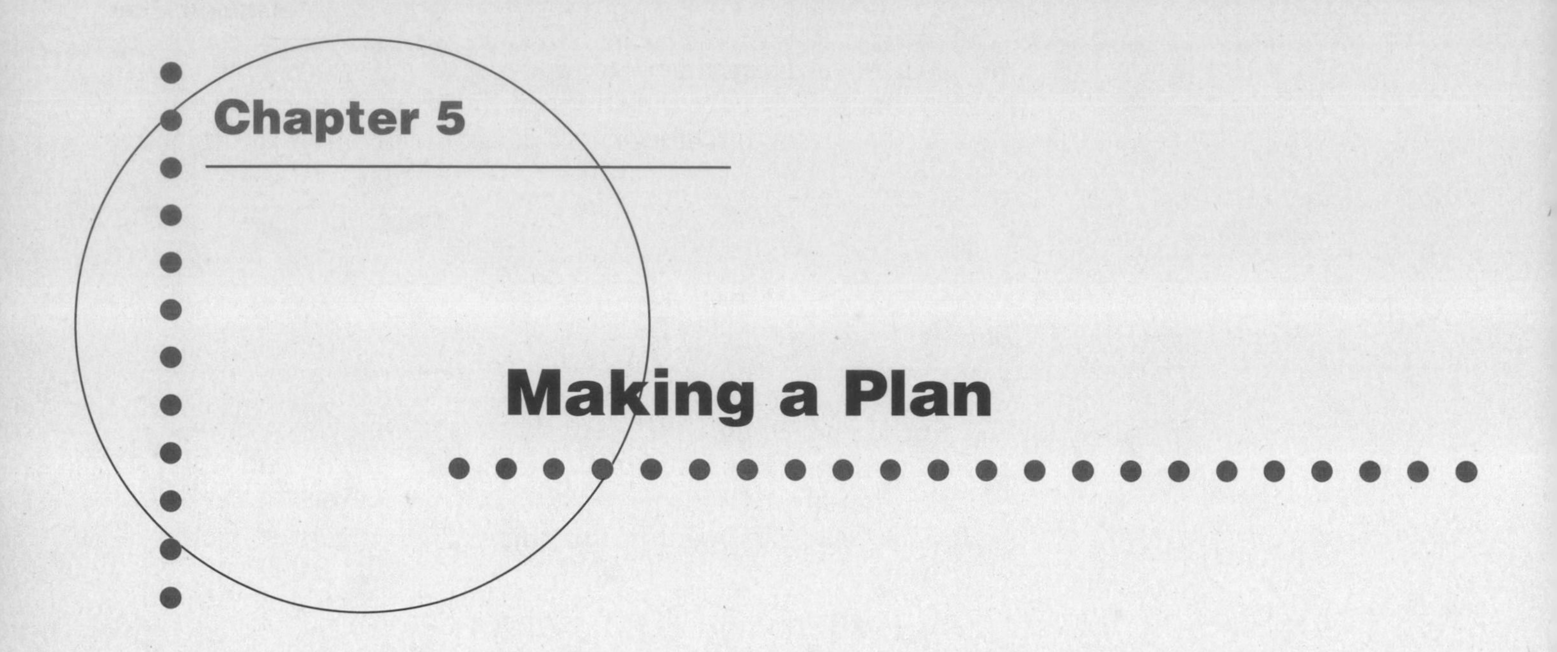

Chapter 5

Making a Plan

CHOOSING THE OPERATION

A carpenter needs 24 nails to put together 1 shelving unit. She wants to build 6 units. How many nails in all will she need?

(1) 4
(2) 18
(3) 30
(4) 144
(5) not enough information is given

What is the correct answer?

How did you decide whether to add, subtract, multiply, or divide?

In the problem above you are given a group (24 nails), and you are asked to find the total in 6 groups. Therefore, you multiply 24 by 6 to find that the answer is **(4) 144**.

Remember that when you are trying to decide how to solve a word problem you should "picture" the problem to understand what is going on. The chart on the next page shows how picturing the problem can help you choose the right operation.

If the problem gives	and you want to	you should
A two or more numbers	put them together in the same group	add
B one number being taken away from another	find out what is left	subtract
C two amounts	find out how much larger one is than the other	subtract
D several items in a group and a number of groups	find the *total* number in all of the groups	multiply
E the cost, weight, or measure of *one* item	find the *total* cost, weight, or measure of many	multiply
F the total number of items and a number of groups	find the number of individual *items* in each group	divide
G a total number of items and the number in each group	find the number of *groups*	divide

Notice the answer choices in the problem on page 70. In many multiple-choice questions some of the incorrect answer choices are based on using the wrong operation. (For example, **(1)** is based on $24 \div 6 = 4$.)

Exercise 1

Directions: Using the ideas on page 71, decide which operation you should use to solve each problem and write the symbol of the operation in the brackets. [+, −, ×, or ÷] Then solve the problem using the correct operation.

Example: Betty ran 5 miles last Thursday. On Tuesday she ran 7 miles. How many more miles did she run on Tuesday than on Thursday?

Operation [—] **Answer:** 7-5 = 2miles

WN

1. When his class took a field trip yesterday, Mr. Rodríguez split the students up into groups of 8. The total class size was 40 students. How many groups were there?

   Operation [ ] **Answer:** ______________

WN

2. The Fontines had 655 bricks to sell. One customer came and purchased 435 of them. How many bricks were left?

   Operation [ ] **Answer:** ______________

WN

3. There were 96 children, 45 men, and 20 women at the festival on Saturday. How many people were there in all?

   Operation [ ] **Answer:** ______________

WN

4. Ted used 4 dozen eggs in his cooking for the senior citizens' home. How many single eggs did he use?

   Operation [ ] **Answer:** ______________

WN

5. How many ounces larger is Can A than Can B shown at right?

   A 15 ounces   B 9 ounces

   Operation [ ] **Answer:** ______________

6. Catherine spent $54.90 at the grocery store last week. This week she spent $50.75. How much more did Catherine spend last week?

   Operation [ ] **Answer:** ______________

D

7. A dry cleaner charges $.80 per shirt for overnight service. If Terry drops off 5 shirts, how much will he have to pay?

Operation [ ] **Answer:** ______________

D

8. The supervisor at Pente Construction paid out a total of $750.50 in wages today. Each worker received $150.10. How many workers were paid?

Operation [ ] **Answer:** ______________

9. Sam is 4 feet, $9\frac{1}{2}$ inches tall, and his older brother Jack is 5 feet, $6\frac{1}{2}$ inches tall. How much taller is Jack?

Operation [ ] **Answer:** ______________

M

10. Tyrone cut the board below into 4 equal pieces. How many centimeters was each piece?

104 cm

Operation [ ] **Answer:** ______________

Answers begin on page 186.

EQUATIONS

> To qualify for the swim team, Jeanette has to swim a total of 325 laps. Jeanette will swim the same number of laps every day over the next 5 days. How many laps will she have to swim daily to qualify?

How do you decide what operation to use to solve this problem?

Try rewording the problem using no numbers, then write a plan to help you solve it.

Here is an example of the problem written without numbers. Decide on a plan for solving it, then write an **equation** (also called a **number sentence**) to organize your plan.

REWORD: Jeanette needs to swim some laps. She wants to swim the same number of laps for several days. How many laps does she need to swim each day?

PLAN: Divide total laps by the number of days to find the number of laps each day.

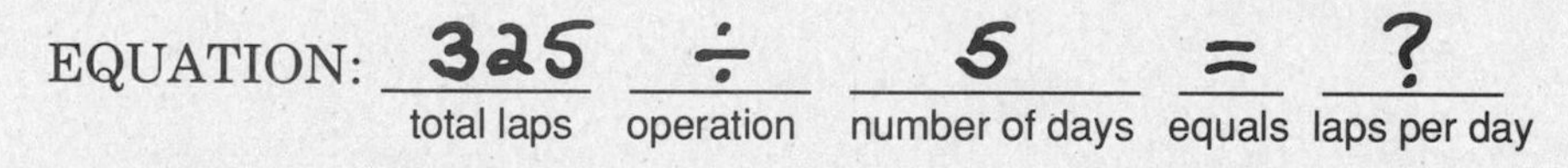

Jeanette will need to swim 325 ÷ 5 or **65 laps** each day.

Here's another example:

The original price of a handmade scarf was $20.00. The designer, however, had to raise the price to $23.50 to cover some unexpected expenses. By how much was the price of the scarf raised?

REWORD: The original price of a scarf was raised. By how much was the price raised?

PLAN: Take the new price and subtract the old price to find the amount the price was raised.

EQUATION: $23.50 (new price) − (operation) $20.00 (old price) = (equals) $3.50 (amount raised)

When you are writing an equation, write the labels with the numbers. This will help you "picture" the problem and see if your answer makes sense.

Exercise 2

Part One

Directions: Choose the correct equation for each of the following problems.

1. Mara babysat for 5 hours on Monday. For 2 of those hours the children were sleeping. How many hours were the children awake?

a) $5 + 2 = 7$ hours **b)** $5 - 2 = 3$ hours

2. Peter has 117 more miles to go on a 220 mile trip. How many miles has he traveled so far?

a) $220 - 117 = 103$ miles **b)** $220 + 117 = 337$ miles

3. Tremaine collected 140 cans on Saturday morning. His sister collected 30 cans more than Tremaine. How many cans did Tremaine's sister collect?

a) $140 - 30 = 110$ cans **b)** $140 + 30 = 170$ cans

D

4. A farmer sells eggs for \$.70 per dozen. In a normal week he sells 35 dozen eggs. How much money does he take in each week from egg sales?

a) $\$.70 \times 35 = \$24.50$ **b)** $\$.70 \div 35 = \$.02$

F

5. The clerk at Hannah's Fabric Store cut this piece of cotton into strips that measured $\frac{1}{3}$ foot wide. How many strips did she get out of the cotton?

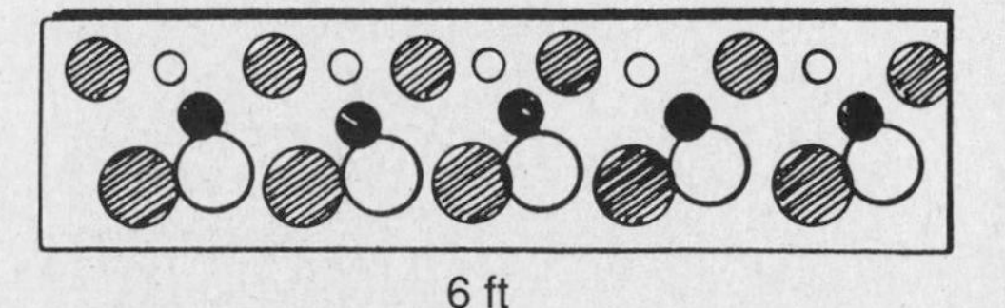

a) $6 \div \frac{1}{3} = 18$ strips **b)** $6 \times \frac{1}{3} = 2$ strips

Part Two

Directions: Write a plan and an equation for each of the following problems. Then solve the problem.

1. Carl works 8 hours each day. If his workweek is 6 days, how many hours does Carl work per week?

   PLAN:

   EQUATION:

2. An average load of laundry at the CleanMart takes 17 minutes to wash and 45 minutes to dry. How long does it take to wash and dry one load?

   PLAN:

   EQUATION:

3. Frances drove 76 miles of the 300-mile distance between her home and her ex-husband's. How many more miles does she need to travel?

   PLAN:

   EQUATION:

4. According to the sign at right, if a customer had 6 shirts dry-cleaned, how much would he pay?

   PLAN:

   EQUATION:

Dry-Cleaning Prices	
Suits →	$9.00
Shirts →	$1.00
Pants →	$3.50
Coats →	$11.00

5. The people attending a nurses' conference were asked to break up into small groups of 15 people. There were 225 people in all at the conference. How many small groups were made?

   PLAN:

   EQUATION:

6. Two families that live on the same block had a yard sale together. One family took in $87 for the day; the other family sold $112 worth of items. How much money did the families earn in all?

PLAN:

EQUATION:

7. A landscaper planted a new tree every 8 feet along one side of this path. How many trees did she plant?

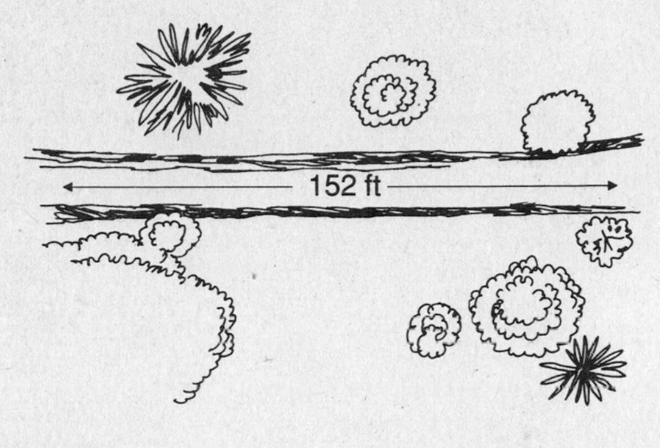

PLAN:

EQUATION:

8. Mary's sandbox holds 3 cubic feet of sand. Right now it is only $\frac{1}{2}$ full. How many cubic feet of sand are in the box?

PLAN:

EQUATION:

9. Fourteen percent of the students at North High are in the vocational program. If there are 800 students at the school, how many are in the program?

PLAN:

EQUATION:

10. How many inches of braid will a tailor need to trim all sides of the square shown at right?

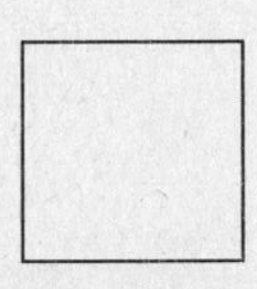

PLAN:

EQUATION:

Answers begin on page 186.

EQUATIONS WITH TWO OPERATIONS

The Elderberry Taxi Service charges $2 for a trip into town and $1 for the return trip. Isabel took 2 round-trips into town this past week. How much did she pay in all?

How many operations do you need to solve the problem?

Your plan for this problem will involve more than one operation. Perhaps you would think something like this:

PLAN: First I should add the costs of the trips to and from town. Then I should multiply the total cost by the number of round-trips. Using what you learned in Chapter 3, you could write:

$$(\$2 + \underset{\text{costs}}{\underset{\uparrow}{\$1}}) \times \underset{\text{trips}}{\underset{\uparrow}{2}} = \mathbf{\$6}$$

This can also be written as $2(\$2 + \$1) = 6$

As you learned in Chapter 3, you should do the operation **inside the parentheses first**. In this case, the parentheses tell you to **add** first, then multiply.

Let's look at another problem with more than one operation.

Ramón started the day with $35.70 in his wallet. He spent $4.50 on breakfast and bus fare, then split the remaining money evenly between his wife and his daughter. How much money did Ramón's wife receive?

PLAN: I should subtract the money Ramón spent from the amount he started with. Then I should divide the remaining amount by the number of people he gave it to.

$$(\underset{\text{money he started with}}{\underset{\uparrow}{\$35.70}} - \underset{\text{amount spent}}{\underset{\uparrow}{\$4.50}}) \div \underset{\text{wife \& daughter}}{\underset{\uparrow}{2}} = \$15.60$$

> **TIP**
>
> **Remember this order of operations:**
>
> 1. Solve parenthesis first.
> 2. Do multiplication and division next.
> 3. Do addition and subtraction last.

Again, see that the parentheses in the equation tell you which operation should be performed first. Look at what would happen if you did the dividing *before* the adding:

RIGHT: (\$35.70 – \$4.50) ÷ 2

= \$31.20 ÷ 2

= **\$15.60**

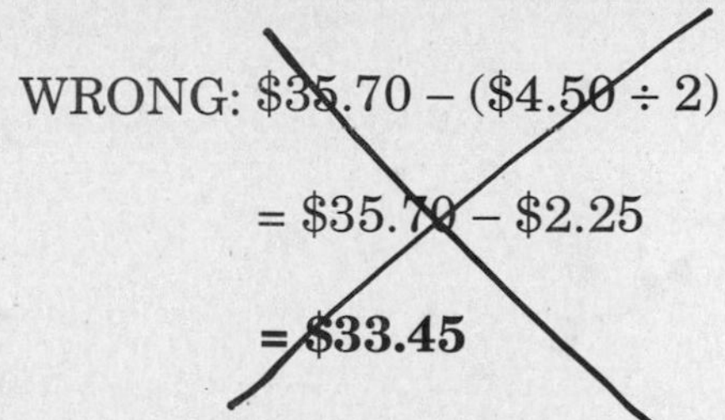

WRONG: \$35.70 – (\$4.50 ÷ 2)

= \$35.70 – \$2.25

= **\$33.45**

When you read a word problem, decide whether it will take one operation or more than one operation to solve. Then write an equation that shows all of the operations.

Exercise 3

Part One

Directions: Choose the correct equation that represents each problem below.

1. Mary has taken 18 sick days in 3 years. On average, how many sick days has she taken each month?

a) 18 ÷ (3 ÷ 12) = 72 **b)** (18 ÷ 3) ÷ 12 = $\frac{1}{2}$

WN

2. An art teacher divided a box of 36 crayons among 4 students. He then added 3 crayons to each student's pile. How many crayons did each student end up with?

a) (36 ÷ 4) + 3 = 12 **b)** 36 ÷ (4 + 3) = 5.14

3. Inés's electric bill came to $64.89 last month. A total of $14.01 of that amount was for services and tax; the remaining amount was for the kilowatt-hours of energy she consumed. If the electric company charges $.06 per kilowatt hour, how many kilowatt hours did Inés consume?

   a) $64.89 – ($14.01 – .06) = 50.94
   b) ($64.89 – $14.01) ÷ .06 = 848

4. A building company is deciding whether or not to purchase two adjacent lots; one lot measures 1.75 acres, and the other measures 2.8 acres. The owner will sell the land for $1,244 per acre. How much would both lots cost?

   a) (1.75 + 2.8) × $1,244 = $5,660.20
   b) 1.75 + (2.8 × $1,244) = $3,484.95

5. A factory worker is responsible for cutting wire tubes like the one shown into 2-inch lengths. If the worker cuts 20 tubes in 1 hour, how many 2-inch lengths will he have at the end of an hour?

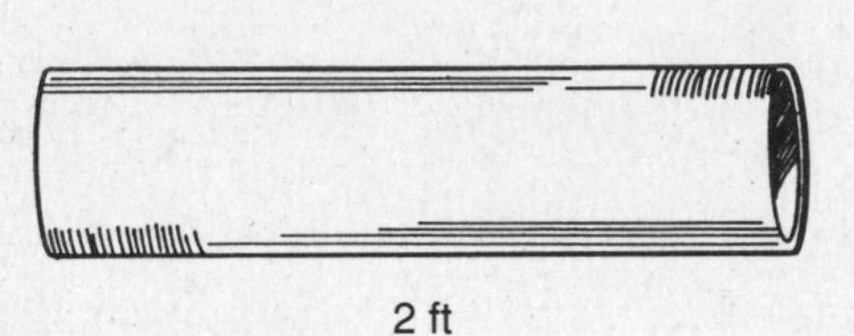

   a) (24 × 20) ÷ 2 = 240
   b) 24 ÷ (20 ÷ 2) = 2.4

Part Two

Directions: Write an equation for each of the following problems. Then solve the problems.

1. Lisa made 3 dozen cupcakes to take to her office on Valentine's Day. She left 14 cupcakes on the second floor, then split the remaining cupcakes among the 22 people who work on the fifth floor. How many cupcakes did each worker on the fifth floor receive?

   EQUATION:

2. Computer World's newest printer can print an average of 400 words per minute. How long would it take for the printer to print Ms. Savage's weekly report if that report is 15 pages long with 200 words per page?

   EQUATION:

3. Mrs. Teel bought 2 swordfish at the local market; 1 weighed 3.5 pounds, and the other weighed 2 pounds. If the fish totaled $19.80 before tax, how much did it cost per pound?

EQUATION:

4. Last week Howard worked $8\frac{1}{2}$ hours on Monday, $4\frac{1}{2}$ hours on Tuesday, and $6\frac{1}{2}$ hours on Friday. If Howard earns $7.10 per hour, how much did he earn that week?

EQUATION:

5. Annie Gordon swam $5\frac{1}{2}$ miles this week. Her brother Paul swam $8\frac{1}{2}$ miles. What is the average number of miles the two swam?

EQUATION:

6. Twenty percent of the pounds lost by the Winners' Weight Loss group were lost during the first month of the program. According to the chart, how many pounds were lost in that month?

Winners' Total Weight Loss	
Member	**Pounds**
Kay Hughes	4
Isabel Perkins	8
Janet Liao	5
Margaret Oates	3
Frank Rodríguez	5

EQUATION:

7. Juan's vegetable garden measures 8 feet by 7 feet. The mulch he plans to buy covers 28 square feet per bag. How many bags does he need to cover the garden?

EQUATION:

Answers begin on page 186.

WRITING EQUATIONS FOR WORD PROBLEMS

Ted had $35 in his wallet one morning. At his office, he collected some money from a bet he had made. When he got home, he counted $73 in his wallet. How much money did Ted collect at the office?

What operation should you use to solve this problem?

How did you decide?

Here is a fact about word problems that you may not have thought of:

IN EVERY WORD PROBLEM, SOMETHING *IS EQUAL TO* SOMETHING ELSE.

In other words, all word problems give you certain numbers and tell you that these numbers **are equal to** other numbers. Knowing this fact will help you write **equations** to solve all kinds of word problems.

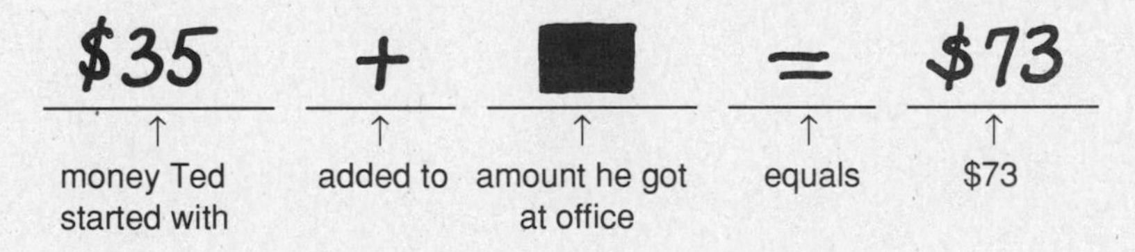

You could represent the equation in several ways:

$$\$35 + \blacksquare = \$73$$

- The box ■ stands for the amount Ted collected at work. You could also write the equation:

$$\$35 + x = \$73$$

- The letter x is often used in equations to stand for *the unknown*—what you are being asked to find.

When you write an equation, you are representing both sides of the equals sign (=) as two equal amounts. Think of this as balancing two sides of a scale.

$$\$35 + x = \$73$$

In other words, what would you have to add to \$35 to "balance" the \$73 on the other side? You probably "knew" that the answer was **\$38**. The next lesson will give you some techniques for solving equations.

Now look at another word problem and equation.

Ann Lih baked some cookies and distributed them to 4 nursing homes in her neighborhood. Each nursing home received 60 cookies. How many cookies did Ann Lih bake in all?

Let x stand for what you do not know—the total number of cookies baked.

What equation could you write for this problem?

x	÷	4	=	60
total cookies	divided by	number of homes	equals	cookies per home

$$x \div 4 = 60$$

Your first step when writing an equation is to make *x* stand for the number you do **not** know in the problem.

What other equations could you write for this problem?

Perhaps you knew from experience that you could multiply to find out what *x* is:

cookies per home ↓ , total cookies ↓

$$60 \times 4 = x$$

↑ number of homes

OR

You may have chosen to divide the total cookies by the number of cookies for each home.

cookies per home ↓

$$x \div 60 = 4$$

↑ total cookies , ↑ number of homes

Each of these equations can be used to find the answer to the word problem. In the next exercise you'll get some practice in writing different equations based on word problems.

Exercise 4

Part One

Directions: For each problem decide what x should stand for—the unknown number you will need to find. Then fill in the blanks of each equation, using x. **Do not solve the equations.**

Example: A total of 17 people showed up for the first day of a computer class. Several of them arrived late. If 11 people came on time, how many were late?

Let x stand for: number who were late

$17 - \underline{x} = 11$ OR $17 - 11 = x$

1. So far this week, Enid has worked 28 hours. How many more hours must she work before she has put in her required 40 hours?

Let x stand for: ______________________

______ $+ x =$ ______ OR ______ $- 28 =$ ______

2. The cost of a phone call Sally made to Doug was $.90 per minute. The total bill for the call was $5.40. For how many minutes did Sally and Doug talk?

Let x stand for: ______________________

$.90 \times$ ______ $=$ ______ OR ______ $\div .90 =$ ______

3. Tania recently spent a total of $44.99 on denim pants for her children. She bought 3 pairs—all at the same price. What was the cost of a single pair of pants?

Let x stand for: ______________________

$3 \times$ ______ $=$ ______ OR $44.99 $\div$ ______ $=$ ______

4. Mrs. Murray plans to use all of a 120-square-foot piece of carpet to cover her laundry room floor. The length of the room is 12 feet. How wide is the room if the piece of carpet is big enough to cover the whole room?

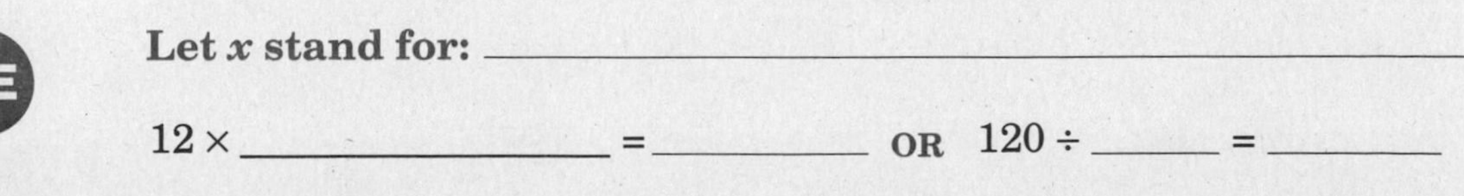

Let x stand for: ______________________

$12 \times$ ______ $=$ ______ OR $120 \div$ ______ $=$ ______

Part Two

Directions: For each problem first decide what x will stand for. Then write **two different** equations using x. **Do not solve the equations.**

Example: Patrick delivered 87 of his newspapers by car. To deliver the remaining 21 papers, he used a wagon. How many newspapers did Patrick deliver?

Let x stand for: total papers delivered

Equation #1: $x - 87 = 21$ **Equation #2:** $21 + 87 = x$

1. An employee of the city subway worked a 35-hour week. If he worked the same number of hours on each of 7 days, how many hours did he work per day?

   Let x stand for: ______________________

   Equation #1: ______________ **Equation #2:** ______________

2. Many of the people in Dora's women's group live outside the city. Of the 29 people in the group only 7 live in the city. How many people in Dora's group are from outside the city?

   Let x stand for: ______________________

   Equation #1: ______________ **Equation #2:** ______________

3. Gary had 80 bunches of daisies to sell to passing cars. The total number of daisies in his cart was 720. How many daisies made up each bunch?

   Let x stand for: ______________________

   Equation #1: ______________ **Equation #2:** ______________

4. Vanessa has $21 to spend on party favors for her son's birthday. Seven boys will be at the party. How much money can Vanessa spend on each boy?

   Let x stand for: ______________________

   Equation #1: ______________ **Equation #2:** ______________

Answers begin on page 187.

SOLVING A WORD PROBLEM EQUATION

A butcher cut up several chickens and divided them into 4-pound packages. He ended up with 9 packages. How many pounds of chicken did the butcher start with?

Write an equation for this word problem.

What is the answer to the problem?

TIP

Do You Remember That . . . ?

- An equation is like a balanced scale?

Whatever you do to one side of an equation you must also do to the other side to keep things equal and balanced.

Suppose you are not sure how to solve this problem. You write an equation to help you solve it. You let x stand for the total pounds of chicken because that is the number you do not know. Your equation would look like this:

$$x \div 4 = 9$$

(number of bags → 9; total pounds → x; pounds per bag → 4)

The following guidelines will help you learn to solve equations such as this one:

1. TO SOLVE ANY EQUATION, YOU MUST GET THE UNKNOWN *(x)* ALONE ON ONE SIDE OF THE EQUAL SIGN.

In the equation you want x to stand alone. In other words you want to get rid of the $\div 4$ that is next to it.

2. USE THE OPPOSITE OPERATION TO GET THE UNKNOWN ALONE.

To get rid of the $\div 4$, you must **multiply** both sides of the equation by 4. Multiplication is the opposite of division.

$$x \div 4 = 9$$
$$x \div 4 \times 4 = 9 \times 4$$

3. SIMPLIFY THE EQUATION.

When you divide by 4 and then multiply by 4, the operations cancel each other out and your unknown stands alone.

$$x \cancel{\div 4 \times 4} = 9 \times 4$$
$$x = 9 \times 4$$
$$x = 36$$

The butcher started with **36 pounds** of chicken.

Take a look at how to solve another word problem using an equation.

> Harold deposited some money in his account yesterday, making a balance of $434.90. If his balance was $325.10 before the deposit, how much did Harold put in his account?

Do You Know Math Operation Opposites . . . ?

- The opposite of adding is subtracting.
- The opposite of subtracting is adding.
- The opposite of multiplying is dividing.
- The opposite of dividing is multiplying.

Write an equation based on this problem.
Think about how you would solve it.

Although there are several different ways to solve this problem, try using this equation:

amount before deposit ↓ total after deposit ↓

$$\$325.10 + x = \$434.90$$

↑ amount of deposit

1. Decide what number needs to be removed to get the unknown alone on one side of the equals (=) sign.	To get x alone in this problem, I need to get rid of the $325.10.
2. Determine what operation is being performed.	The 325.10 is being *added* to x.
3. Perform the **opposite** operation to **BOTH** sides of the equation.	The opposite of adding is subtracting. $325.10 - 325.10 + x = 434.90 - 325.10$
4. This opposite operation cancels out the numbers next to the unknown. You are left with a new equation with x alone.	$x = \$434.90 - 325.10$ $x =$ **$109.80**

"Switching Places" in Equations

Let's take a look at one more example of equations with word problems. This one is a little tricky.

> The Sidfield Plant employed 1,300 people in 1988. Due to a slow economy, many people were laid off in early 1989, leaving Sidfield with only 590 workers. How many people were laid off?

Write an equation to solve this problem.
Can you think of more than one equation?

Here is one equation you could write for this problem:

$$\underset{\text{1988 total}}{1{,}300} - \underset{\text{number laid off}}{x} = \underset{\text{1989 total}}{590}$$

Because there is a minus sign (–) in front of the unknown, this is a more difficult equation to solve. You would need to perform several more steps to get the x by itself. Now look at another equation for the same problem.

$$\underset{\text{1988 total}}{1{,}300} - \underset{\text{1989 total}}{590} = \underset{\text{number laid off}}{x}$$

The two equations—$1{,}300 - x = 590$ and $1{,}300 - 590 = x$—have the same solution: 710. The second one is easier to solve because the x is already alone on one side of the equal sign.

If an equation has an unknown with either a minus (–) or a division (÷) sign in front of it, rewrite the equation as shown here:

switch

Example: $36 \div x = 9$ becomes $36 \div 9 = x$

$36 - x = 4$ becomes $36 - 4 = x$

switch

Exercise 5

Part One

Directions: Use the steps for solving equations to find x.

1. $37 + x = 91$

2. $x \div 13 = 11$

3. $144 \div 4 = x$

4. $x - 194 = 19$

5. $80 - x = 14$

6. $14 \times x = 42$

7. $144 \div x = 12$

8. $27 + x = 120$

9. $x \div 12 = 9$

10. $105 \div x = 7$

Part Two

Directions: For each word problem write an equation and solve it.

Example: Vera paid for her groceries with a $20 bill. She received $3.14 in change. How much did Vera's groceries cost?

Equation: $\$20 - x = \$3.14$ OR $\$20 - \$3.14 = x$

Solve: $\$20 - \$3.14 = \$16.86$

1. When they drew up their will, Nathan and his wife decided to divide their collection of 96 rare coins evenly among their children. If each child is to receive 32 coins, how many children do Nathan and his wife have?

2. A stockperson added 12 pounds of food to a crate, making a total of 60 pounds of food. How many pounds of food were in the crate before the 12 pounds were added?

3. Several committees were formed to plan for the March Against Drugs. Each committee had 15 people. If a total of 225 people were in committees, how many committees were formed?

4. Juan withdrew $145 from his savings account, leaving a balance of $1,004. How much money was in Juan's account before the withdrawal?

5. Hal needed a total of 360 brownies to serve at the club banquet. Each pan in the kitchen held 40 brownies. How many pans did Hal need to fill?

Answers begin on page 187.

WRITING PROPORTIONS FOR WORD PROBLEMS

In a quality-control check, Rita found that for every 8 usable cups the molding machine was producing 3 rejects. If in 1 hour the machine produced 135 rejects, how many usable cups did it produce?

Can you write an equation to solve this problem?

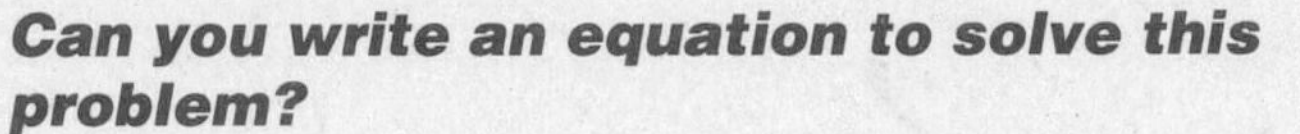

You learned that in every word problem something *is equal to* something else. But in this problem the equality is hard to see at first. What is equal to what?

In this problem two things are being *compared*. The problem is comparing good cups to rejected cups. One way to compare is to show a **ratio**.

$\frac{3}{8}$ ← rejects / ← usable cups

The ratio is 3 rejects for every 8 usable cups.

The ratio could be written as "3 to 8" or 3:8 or $\frac{3}{8}$. To use ratios to solve word problems you will be using the fraction form: $\frac{3}{8}$.

You should remember that to write an equation for any word problem your first step is always to decide what x, the unknown, will stand for.

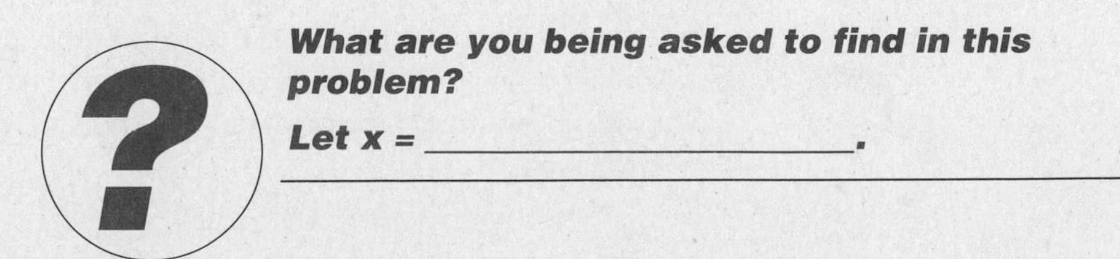

What are you being asked to find in this problem?

Let x = ______________________.

You're right if you let x stand for the **number of usable cups**.

To write an equation you need to set up a proportion. A **proportion** is made up of two equal ratios.

$$\text{rejects} \rightarrow \frac{3}{8} \leftarrow \text{usable cups} \quad = \quad \frac{135}{x} \quad \begin{matrix}\leftarrow \text{rejects} \\ \leftarrow \text{usable cups}\end{matrix}$$

Notice that in a proportion the same category, or label, goes on the top of each ratio (rejects), and the same category goes on the bottom (usable cups).

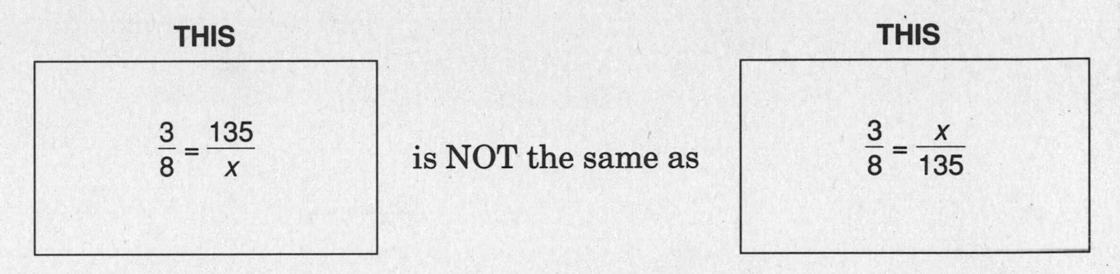

Later you'll learn how to solve proportion equations. For now, use the next exercise to practice setting up proportions.

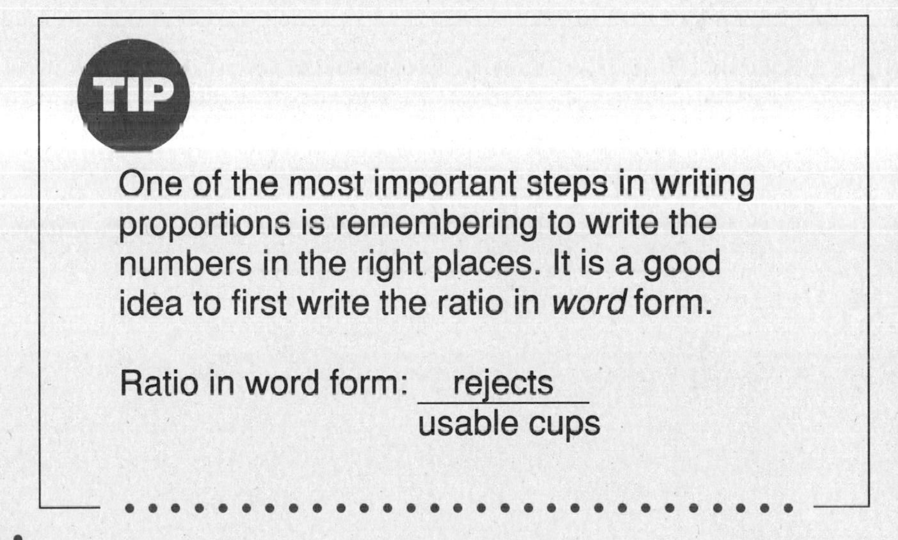

One of the most important steps in writing proportions is remembering to write the numbers in the right places. It is a good idea to first write the ratio in *word* form.

Ratio in word form: $\frac{\text{rejects}}{\text{usable cups}}$

Exercise 6

Part One

Directions: Read each of the following problems, then fill in the missing words and numbers in the proportion. Always use x to stand for the unknown number. **Do not solve the problems.**

Example: At Solly's Fresh Produce, bruised apples are sold for \$.90 per dozen. At this price, how much will 26 apples cost?

Compare: apples / cost

$$\frac{12}{\$.90} = \frac{26}{x}$$

fill in

WN

1. The Beautiful Bus tour covered 900 miles in 16 hours. If it continues at the same rate, how many hours will it take to cover the remaining 225 miles?

Compare: $\frac{\text{miles}}{\_\_\_\_\_} \qquad \frac{900}{16} = \frac{\_\_\_\_\_}{\_\_\_\_\_}$ ← miles

label → ← unknown

2. The scale of miles on a road map shows that 1 inch represents 300 miles. Todd estimated the road-map distance from Lawton to Dawson to be about 3 inches. How many miles is it from Lawton to Dawson?

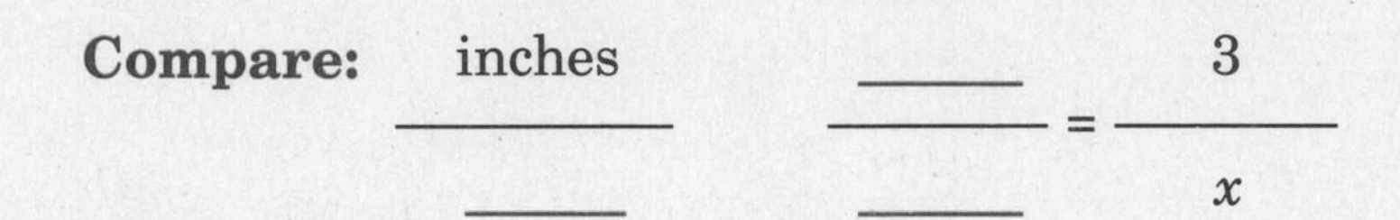

3. Mariel's cake recipe calls for 7 ounces of chocolate to every 3 ounces of sugar. If she uses 21 ounces of chocolate, how many ounces of sugar will she need?

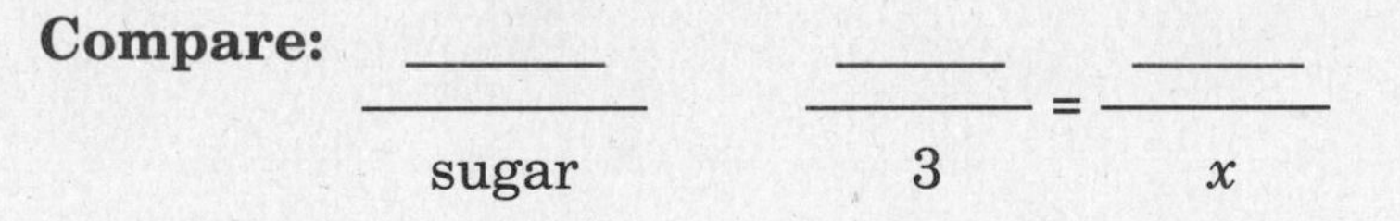

4. If a 20-ounce package of chicken costs $4.39, how many ounces is a package that costs $5.28?

Part Two

Directions: For each of the following word problems, first circle the two things that are being compared. Then set up a proportion using x for the unknown number. **Do not solve the problems.**

Example: For every 8 boxes of greeting cards that Melanie sells she receives 1 free movie pass. How many boxes of cards will Melanie need to sell to earn 4 movie passes?

$$\frac{\text{boxes of cards}}{\text{movie passes}} \qquad \frac{8}{1} = \frac{x}{4}$$

WN

1. At the Pasta House, a strand of spaghetti must be 7 inches long to be packaged and sold. On a good day the workers produce 15 usable strands for every 2 strands that can't be sold. If in one day they had to throw out 1,000 rejects, how many good strands could they package and sell?

WN

2. Mrs. Gutiérrez filled her gas tank and drove 352 miles on 11 gallons. At this rate, how many more miles can she drive on the remaining 4 gallons in her tank?

F

3. To get the right color of paint for Tanisha's apartment, the landlord must mix 2 gallons of blue with every 3 gallons of white. If the landlord used a total of $5\frac{1}{2}$ gallons of blue paint, how many gallons of white did he use?

M

4. A pump can drain 4,800 liters of water from a pond in 50 minutes. How many liters can it drain in an hour and a half?

GE

5. Nancy took a 3-inch-wide by 5-inch-long photograph to a specialty camera shop. She wanted it enlarged to fit exactly in a 20-inch-long frame. How wide will the new picture be?

Answers begin on page 188.

SOLVING PROPORTION EQUATIONS

On his test, Willie was told that he answered 9 questions correctly for every 2 questions he got wrong. When he looked at his paper, Willie saw that he got 45 questions correct. How many questions did he get wrong?

Set up a proportion for this word problem.

How can you solve this equation?

You may have written a proportion like this:

$$\frac{\text{correct}}{\text{wrong}} \begin{matrix} \rightarrow \\ \rightarrow \end{matrix} \frac{9}{2} = \frac{45}{x}$$

Or perhaps you chose to put the number of wrong answers on top:

$$\frac{\text{wrong}}{\text{correct}} \begin{matrix} \rightarrow \\ \rightarrow \end{matrix} \frac{2}{9} = \frac{x}{45}$$

It doesn't matter which way you choose, as long as **the same categories are on top and the same categories are on bottom.**

To solve a proportion equation, remember this:

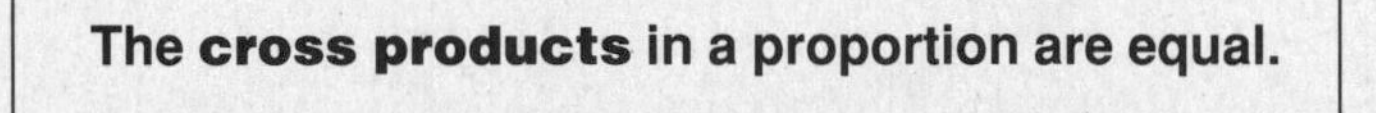

The cross products in a proportion are equal.

For example:

$$\frac{9}{2} \times \frac{45}{x}$$

Cross products are: $9 \times x$ and 2×45

Step 1: Multiply the cross products.

$$9 \times x = 2 \times 45$$
$$9x = 90$$

Step 2: Divide to get the x alone: $x = 90 \div 9$

$x = 10$

Willie got **10 questions** wrong on his test.

Solve the problem below by following the steps.

A car at Hughes Motor Sales cost $3,600 with a tax of $120. At this same rate, how much tax would a customer pay on a car that costs $4,500?

1. Decide what numbers are being compared.	The problem is comparing *car cost* to *tax*.
2. Decide what x will stand for.	The unknown is the tax on a $4,500 car.
3. Set up a proportion with the same categories on top.	$\frac{\text{tax}}{\text{cost}} \quad \frac{120}{3{,}600} = \frac{x}{4{,}500}$
4. Write an equation using cross products.	$120 \times 4{,}500 = 3{,}600 \times x$
5. Solve the equation.	$540{,}000 = 3{,}600 \times x$ $540{,}000 \div 3{,}600 = x$ $150 = x$

The tax on a $4,500 car would be **$150**.

Exercise 7

Part One

Directions: Cross multiply and divide to solve the following proportion equations for x.

1. $\frac{x}{5} = \frac{4}{10}$
2. $\frac{8}{4} = \frac{60}{x}$
3. $\frac{x}{16} = \frac{200}{80}$
4. $\frac{9}{x} = \frac{6}{4}$
5. $\frac{13}{5} = \frac{x}{10}$
6. $\frac{8}{3} = \frac{24}{x}$

Part Two

Directions: Go back to the proportion equations you wrote in Part Two of Exercise 6 on page 93. Solve for x.

Answers begin on page 188.

WHEN CAN YOU USE A PROPORTION?

The assembly line on which Misha works can wrap 93 boxes of pencils in 2 hours. How many hours will it take for this line to wrap an order of 465 boxes?

Can you use a proportion to solve this problem?

How do you know?

When you are deciding how to solve a word problem, use this test to find out whether or not you can use a proportion equation:

1. Find two things in the problem that are being compared. • • • • • *boxes* and *hours*

2. Make a chart and fill in the two things being compared. • •

boxes		
hours		

3. Now fill in any values given in the problem. Use x for the unknown. • • • •

boxes	93	465
hours	2	x

4. If you can fill in three values with the fourth value missing, you can write and solve a proportion equation. • •

$93x = 465 \times 2$
$x \times 93 = 930$
$x = 930 \div 93$
$\boldsymbol{x = 10}$ **hours**

Now try the same steps above with this problem:

Admission to the flea market is $.75 for children and $1.25 for adults. What would the admission costs be for the Sobala family of two adults and one child?

1. Find two things in the problem that are being compared. • • • • • • *cost for adults* and *cost for children*

2. Make a chart and fill in the two things being compared. • •

adults $		
children $		

3. Now fill in any values given in the problem. • • • •

adults $	$1.25	
children $	$ .75	

4. If you can fill in three values with the fourth value missing, you can write and solve a proportion equation. • • • You *CANNOT* fill in any more of the chart.

TIP

When you are given three numbers in a problem, and two of them have the same label (*dollars, boxes, feet, hours, dogs, shoes,* etc.), try writing a proportion equation. While it may not always work, it is worth a try.

For the second problem, the proportion method will not work. You must find another way to solve the problem.

As you get more and more experienced with word problems, you will be able to see more quickly when to use a proportion.

Exercise 8

Part One

Directions: Use the steps above to decide which of the problems can be solved using a proportion. Fill in as much of the chart with labels and numbers as you can. Use x to stand for the unknown. Then check YES or NO. **Do not solve the problems.**

1. At Berry Book Sales there is one part-time employee for every 6 full-time employees. If there are 13 part-time employees in all, how many full-time employees are there?

label →	part-time		
label →	full-time		

Can this problem be solved using a proportion? YES ____ NO ____

D

2. Kathryn pays $15 to have 40 clean diapers delivered to her house. At this rate, what would she pay for 100 diapers?

label →			
label →			

Can this problem be solved using a proportion? YES ____ NO ____

F

3. Wendy's fruit salad recipe calls for 2 cups of yogurt for every 5 pounds of fruit used. Wendy has only $1\frac{1}{2}$ cups of yogurt. How much fruit should she use?

label →			
label →			

Can this problem be solved using a proportion? YES ____ NO ____

4. During its busy season, the Oak Mall serves about 1,600 people each week. During the slow season, it serves about 70% of this number. How many people are served weekly in the slow season?

label →			
label →			

Can this problem be solved using a proportion? YES ____ NO ____

5. Reed's father built a tree house that was 5 feet high, 5 feet wide, and 6 feet long. How many cubic feet was Reed's tree house?

label →			
label →			

Can this problem be solved using a proportion? YES ____ NO ____

Part Two

Directions: Go back and find answers to the problems in Part One that **can** be solved by using a proportion.

Answers begin on page 189.

DRAWING A PICTURE

An electrician needed to cut several short pieces of wire from a spool that held 130 inches of wire. He wanted each piece to measure 5 inches long. How many pieces can he get from this spool?

Do you multiply or divide to find the answer?

How did you decide?

Sometimes you may know right away which operation to use to solve a word problem. Other times you may write a number sentence or equation that helps you decide. Here is another strategy that can help if you get stuck: you can draw a picture.

Here's a drawing of what is going on in the problem above:

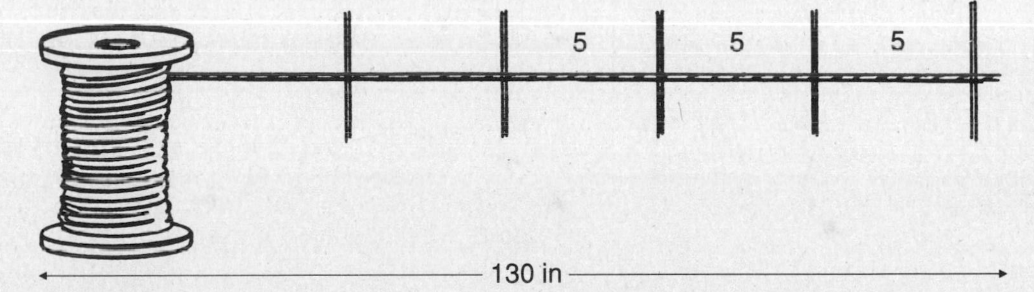

DID YOU KNOW . . . ?

- You don't have to be a very good artist to draw a simple sketch of a problem.
- You are the only one who has to be able to understand your drawing—it has to make sense only to **you**.
- A drawing is simply a tool that helps you visualize what is happening in a problem.

With the drawing above you might find it easier to see that you should **divide** 130 by 5 to get your answer.

$130 \div 5 =$ **26 pieces**

Let's look at another problem.

> The General K Company decided to make its new cereal box $\frac{1}{4}$ larger than its old box. If the old box held 32 ounces, how many ounces will the new box hold?

Draw a picture of what is happening in this problem.

What operations will you need to use to solve it?

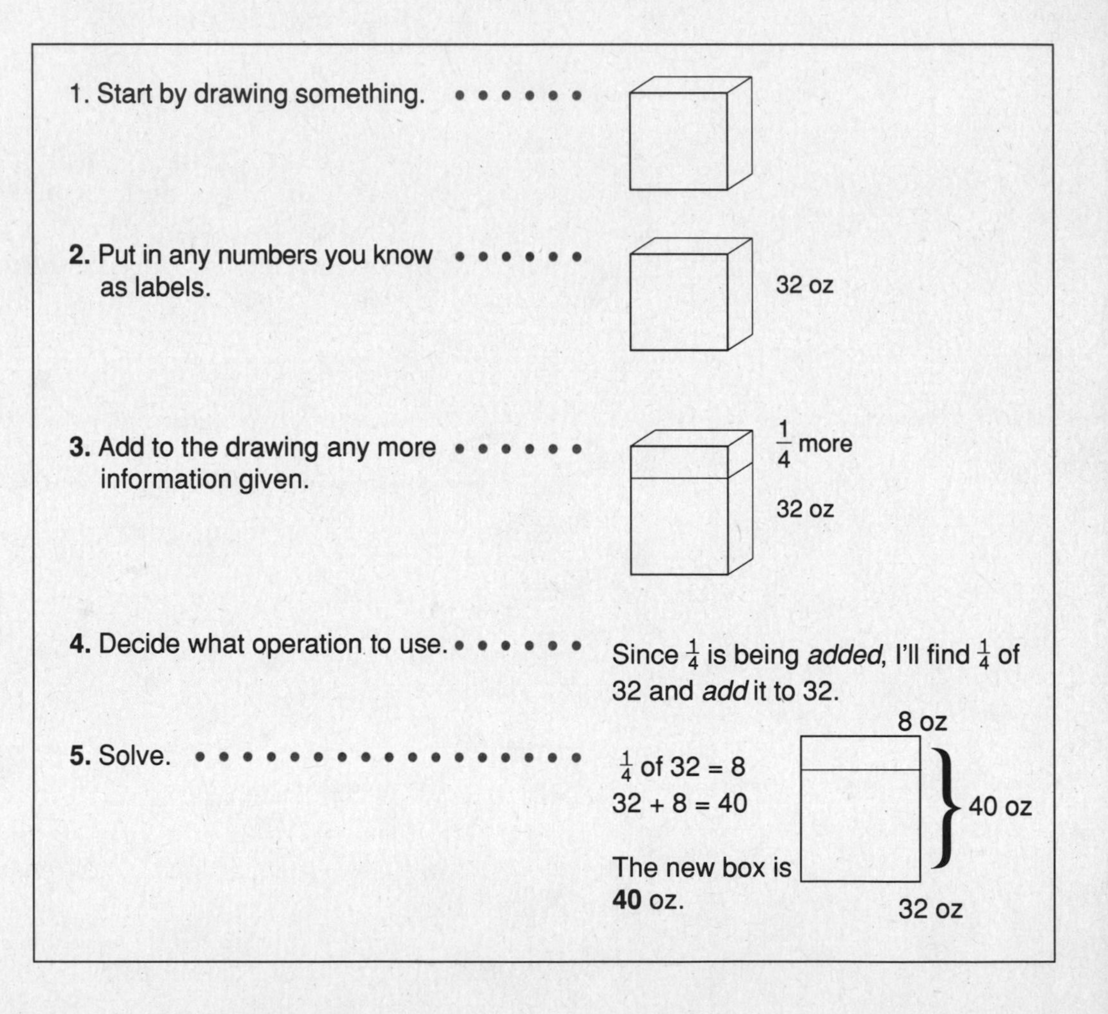

Exercise 9

Part One

Directions: Each of the following problems has a picture with it. Use the picture to solve the problem.

Example: By 7:00 P.M. Tom's rain barrel had collected 2 inches of rain. By 9:00 P.M. the barrel contained 3 inches. How many inches had been collected between 7:00 and 9:00 that night?

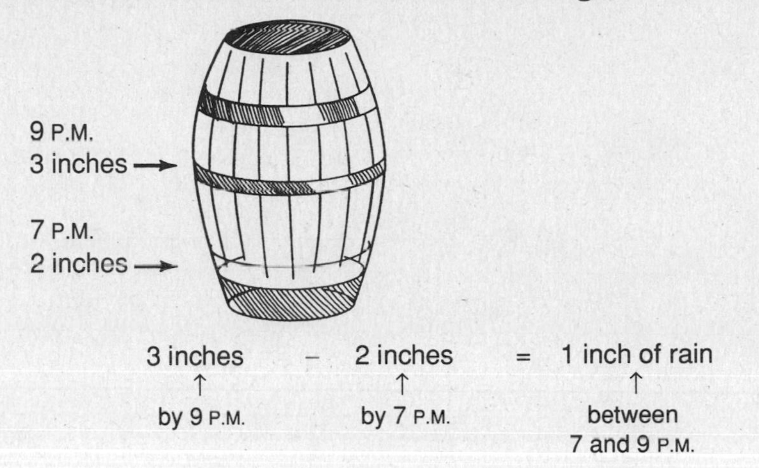

3 inches (by 9 P.M.) − 2 inches (by 7 P.M.) = 1 inch of rain (between 7 and 9 P.M.)

1. A political organization has to get 960 signatures on its petition. If each petition holds 24 signatures, how many petitions must be filled?

2. A worker spent the day driving steel posts into the ground. Each post is set into a hole that is $2\frac{1}{2}$ feet deep. If the post measures 8 feet tall, what is the aboveground height of the post?

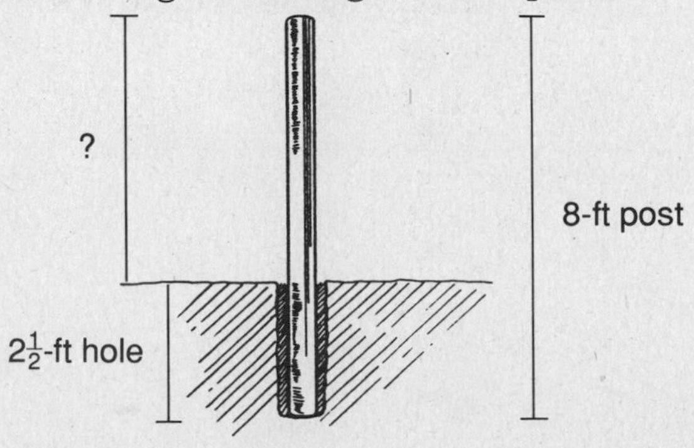

3. The soup-kitchen volunteers need to use up 99 sticks of margarine before they spoil. If the bread recipe they are using requires 3 sticks per loaf, how many loaves can they make?

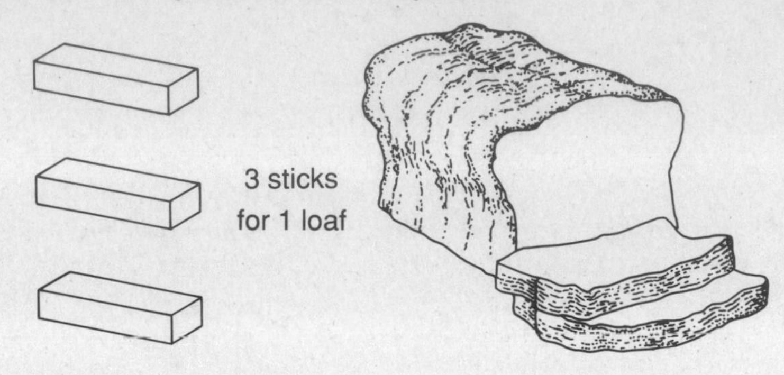

4. Of the 27 miles between home and work, Nigel has traveled halfway. How many more miles does he have to go?

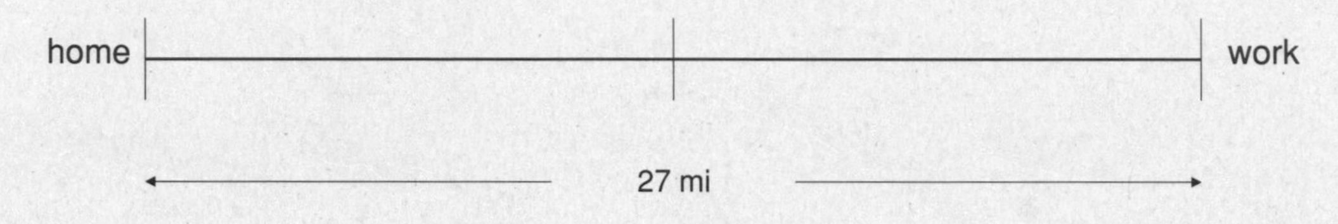

5. A seamstress shortened a skirt from $34\frac{1}{2}$ inches to $31\frac{1}{4}$ inches. How many inches did she take off the skirt?

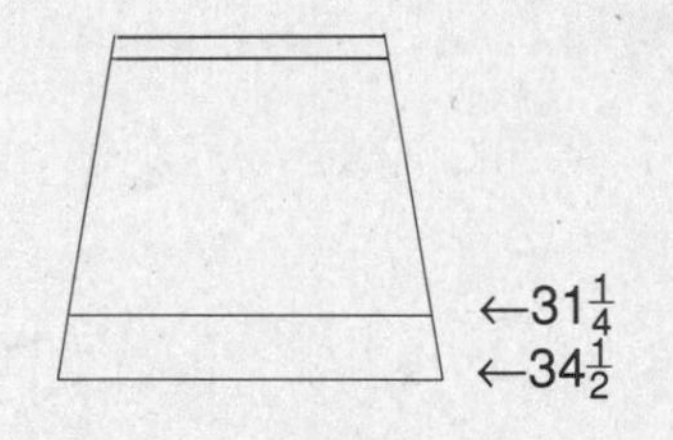

6. A warehouse employee makes a stack of packing cartons that is 10 cartons high, 5 cartons wide, and 6 cartons deep. If each carton holds 8 cans, how many cans are in the stack?

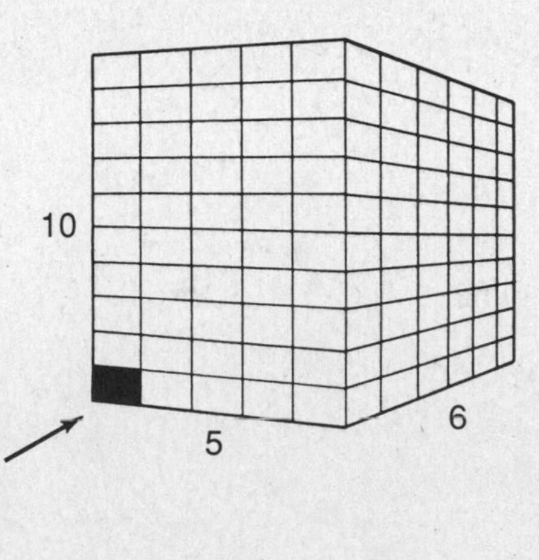

Part Two

Directions: Solve the following word problems. Make a drawing to "picture" the problem.

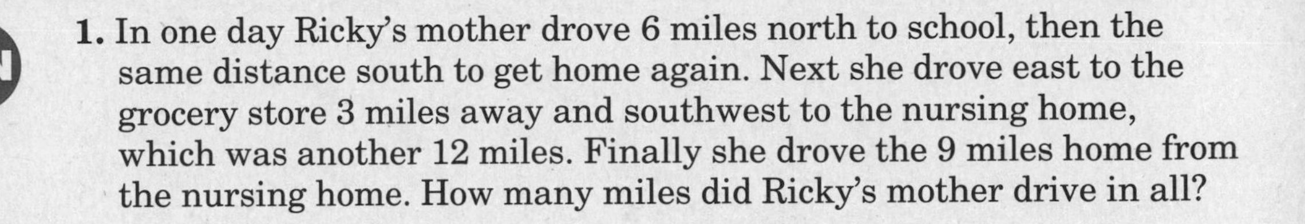

WN

1. In one day Ricky's mother drove 6 miles north to school, then the same distance south to get home again. Next she drove east to the grocery store 3 miles away and southwest to the nursing home, which was another 12 miles. Finally she drove the 9 miles home from the nursing home. How many miles did Ricky's mother drive in all?

2. Mrs. Antony knows that she needs $\frac{1}{2}$ can of soup for every pound of beef that she wants to stew. If she has 12 cans of soup in her cupboard, how many pounds of beef can she stew?

3. Lee knew that she had to plow a certain amount each day to get her fields prepared for planting. The first day she plowed 16 acres, and the next day she plowed 25% of what she had on the first day. How many acres did she plow altogether?

4. To buy the correct amount of paint for his room, Fred has to figure out the area of the wall space. Each wall is square and measures 10 feet per side. What is the total area of all 4 walls?

5. Estella makes hair scarves for extra money. She wants to put sequined trim along the edges of a triangular kerchief that is 2 feet on each of 2 sides and 2 feet, 4 inches on the third side. How much trim does she need?

6. A rectangular field is twice as long as it is wide. If the field is 85 yards wide, how long is the field?

Answers begin on page 189.

MIXED REVIEW

Directions: Solve the following word problems. Use equations, proportions, pictures, and charts to help yourself.

WN

1. A maintenance man uses 6 gallons of white paint and 4 cans of varnish for every apartment he remodels. He has 2 apartments left to remodel. If he has only 2 gallons of white paint on hand, how many more will he need to buy to finish both apartments?

(1) 18
(2) 10
(3) 8
(4) 4
(5) not enough information is given

WN

2. A pharmacy worker had 250 ml of an antibiotic to mix with 400 ml of a flavored syrup. This mixture was then to be divided into small jars containing 50 ml. Which of the following expressions shows the number of jars the pharmacy worker could fill?

(1) $(250 \times 400) \times 50$
(2) $\frac{250 \times 400}{50}$
(3) $\frac{400 - 250}{50}$
(4) $400 - \frac{250}{50}$
(5) $\frac{400 + 250}{50}$

WN

3. For every 2 ounces of infant formula concentrate, Veralee needs 6 ounces of sterilized water to feed her sick baby. If the can of concentrate contains 16 ounces, how many ounces of water will Veralee need?

(1) 8
(2) 32
(3) 48
(4) 96
(5) 108

Temperature Patterns
8-6 through 8-12
Edgarton, Mass.

100°
90°
80°
70°
60°
50°

S M T W T F S

4. According to the graph above, what was the average temperature for the week in Edgarton?

(1) 60°
(2) 65°
(3) 70°
(4) 75°
(5) 525°

F

5. The map shows the route that Mrs. Nguyen takes when it is her turn for the car pool. If the total distance she travels from her home to the school is $7\frac{1}{2}$ miles, what is the distance from Theo's house to the school?

Mrs. Nguyen's Home
Mrs. Nguyen picks up Theo at his house
$3\frac{3}{4}$ miles
School

(1) $8\frac{1}{4}$
(2) $6\frac{1}{2}$
(3) $4\frac{1}{2}$
(4) $3\frac{3}{4}$
(5) 2

P

6. In 1988 the average American watched 49.5 hours of television per week. Approximately 30% of this time is taken up by advertising. About how many hours per week is the average American watching commercials on TV?

(1) 14.85
(2) 16.5
(3) 17.85
(4) 19.5
(5) 165

Questions 7–8 refer to the following information.

Rachel and Martha are purchasing a home together. Because Martha earns more money than Rachel, they have agreed that Martha will pay 60% of the mortgage payment and Rachel will pay 40%. They are equally splitting $12,000 as a down payment, and their monthly mortgage payments will be $750. In addition, they estimate that their utility bills will total $180 each month. Since Rachel works at home, she will pay for all utilities.

7. Martha had $13,350 in her savings account. After she withdraws her share of the down payment, how much will she have remaining in the account?

(1) $1,350
(2) $6,000
(3) $7,350
(4) $12,000
(5) not enough information is given

8. How much money will Rachel pay each month, including mortgage and utilities?

(1) $72
(2) $300
(3) $372
(4) $480
(5) $4,980

Answers begin on page 190.

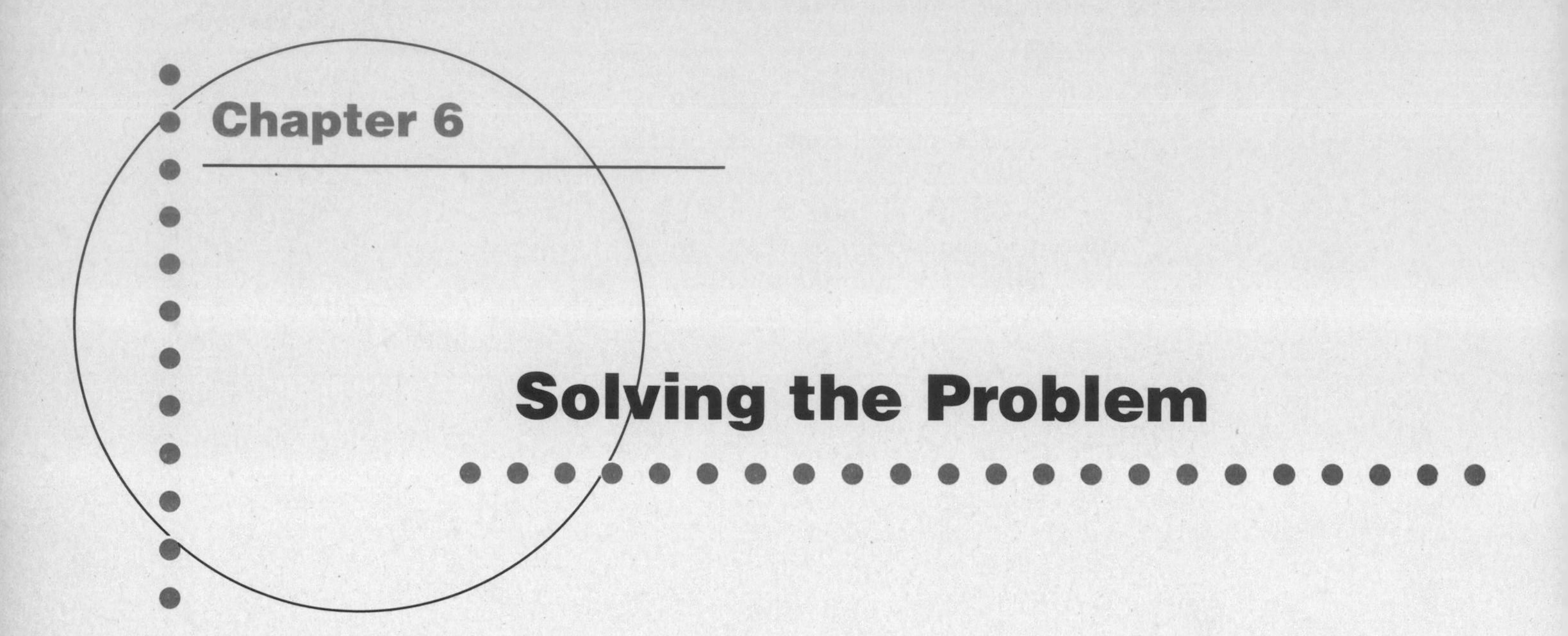

Chapter 6

Solving the Problem

KEEPING ORGANIZED

Jan works in a day-care center 20 hours a week and earns $6 per hour there. At night she works as a waitress in a restaurant, where she earns twice that amount each week. How much money does Jan earn altogether in a week?

(1) $20
(2) $120
(3) $240
(4) $360
(5) not enough information is given

What do you need to do to get the right answer to this problem?

Use the space below to solve the problem. When you are finished, close this book and wait five minutes before you open it again.

Have you waited five minutes? If so, look at your work in the space on page 106 and do the following:

- Circle the number of hours Jan works each week at the day-care center.
- Underline the amount of money she earns weekly at the day-care center.
- Put a box around the amount of money Jan earns weekly as a waitress.
- Put a star next to your answer to the problem.

Was it easy for you to find all the numbers? Or did you have to search around the page and rethink the problem in your head?

Believe it or not, many mistakes are made because people do not work in an organized way. Look at the following solution to the problem above, and go through the steps listed. Notice how easy it is to follow the work.

Day care	Waitress	
\$6 per hour × 20 hours \$120 each week	\$120 × 2 \$240 each week	\$240 +\$120 \$360 total

If you are getting the right answer most of the time, you have probably found a process that works well for you. However, if you think you should be getting the right answer more often, and you are unsure about what you are doing wrong, keep in mind the following.

TO STAY ON THE PATH TO THE RIGHT ANSWER . . .

1. Be sure to line up your numbers neatly.

THIS		IS NOT THIS	
	\$240 +\$120 \$360		\$240 + \$120 \$2,520

2. Keep each step separate and number the steps if it helps.

THIS			IS BETTER THAN THIS
	Step 1:	\$ 6.00 × 20 \$ 120.00	\$6.00 ×20 120
	Step 2:	\$ 120 × 2 \$ 240	120 2 240

TIP

Working in an organized way won't always ensure the right answer, but sloppy and careless work will almost always lead to mistakes.

3. Use labels whenever possible.

THIS WILL NEVER CONFUSE YOU

```
  $ 240 (waitress)
+ $ 120 (day care)
  $ 360 (total)
```

AS THIS WILL

```
  240 (hours?? days?? day care??)
+ 120 (dollars?? hours?? waitressing??)
  360 (hours?? day care?? waitressing??)
```

4. Make sure that the numbers you have written are the same as the numbers in the problem.

THIS $6 per hour, 20 hours

CAN BE COPIED INCORRECTLY AS $20 per hour, 6 hours

Exercise 1

Directions: First solve each problem below using any hints from above that you think might help you. Then read the student's solution that follows each problem. Find the mistake that led to the incorrect answer.

Example: Minni started a business as a carpenter. On weekdays she works 9 hours a day, and she works 4 hours on Saturdays. How many hours does she work per week?

a) Correct Solution

```
  9 hours          45 hours
× 5 days         +  4 hours
 45 hours          49 hours
```

b) Incorrect Solution. What is the mistake in this solution?

```
  9 hours
+ 4 hours
 13 hours
```

The student forgot to multiply the 9 hours by the 5 weekdays.

1. The chart below shows the number of hits the Billtown baseball team got in last week's games. What is the average number of hits for the best 3 hitters?

PLAYER	HITS
Fernando Benson	12
Jim Ryan	1
David Burns	4
Tyson Jones	11
Lewis Caviness	8
Doug Mills	10
John McCarthy	9
Steve Kiernan	9

a) Your solution:

b) What is the mistake in this solution?

12 + 1 + 4 + 11 + 8 + 10 + 9 + 9 = 64

64 ÷ 8 = 8 hits

The mistake is

2. Michael, the neighborhood handyman, mowed 4 lawns, fixed 2 clogged drains, and worked for 8 hours on a remodeling project this week. If each lawn took him $1\frac{1}{2}$ hours to mow and each drain took 1 hour to fix, how many hours in all did he work?

a) Your solution:

b) What is the mistake in this solution?

Step 1: $4 \times 1\frac{1}{2} = 6$ hours mowing

Step 2: $2 \times 1 = 2$ hours fixing drains

Step 3: $6 + 2 = 8$ hours

Answers begin on page 191.

ESTIMATING THE ANSWER

Tara waitressed Saturday and earned $15.75 in tips. The previous Saturday she earned 3 times that much. How much money did Tara earn on the previous Saturday?

(1) $5.25 **(2)** $12.75 **(3)** $31.50 **(4)** $47.25 **(5)** $57.20

Without doing any computation, decide which of the answer choices above might be the right answer.

How did you decide?

Estimation is a useful tool. You can estimate "in your head" or with paper and pencil to:

- find an appropriate answer
- check to see if your answer makes sense

When you **estimate** an answer to a problem, you get a rough idea of the correct answer by rounding off the numbers in the problem to make them easier to work with. Here are some steps to follow:

Step	Work
1. Try an estimated solution.	($15.75 is about $16) $16 × 3 = $48 (estimated answer)
2. Choose the answer choice closest to your estimate.	Choice (4) **$47.25** is closest to $48.
3. Check your estimate with the original numbers.	$15.75 × 3 = **$47.25** (real answer)

On multiple-choice tests, an estimate helps you choose the correct answer quickly.

You can indicate "is approximately equal to" with the sign ≈. In the problem above, $47.25 ≈ $48.

Exercise 2

Directions: For each problem below, first estimate an answer. Then use your estimate to choose from the list of answers. Then find an exact answer and compare it to your estimate.

Example: An assembly-line worker can complete 864 circuit boards in 24 hours. At this rate, how many boards can she complete in 1 hour?

(1) 28 (2) 30 (3) 36 (4) 360 (5) 400

a) Estimate: 864 → 900 circuit boards   900 ÷ 20 = 45
24 → 20 hours

b) Answer choice: (3) **36** 36 is closest to 45.

c) Exact solution: 864 ÷ 24 = 36 This verifies the estimate.

WN

1. Harold weighs 68 pounds more than his mother, who weighs 41 pounds more than Harold's sister. If Harold weighs 211 pounds, what does his sister weigh?

(1) 170 (2) 143 (3) 114 (4) 102 (5) 62

a) Estimate: 68 → ____
41 → ____
211 → ____

b) Answer choice:

c) Exact solution:

2. Hamburger meat is on sale at Harold's Market for $1.48 per pound. How much would a package weighing 3.25 pounds cost?

(1) $2.20 (2) $4.81 (3) $5.75 (4) $8.93 (5) not enough information is given

a) Estimate: **b) Answer choice:** **c) Exact solution:**

D

3. Premium gas at Fred's Fill-Up Station costs $1.86 per gallon. How much did Doreen pay for 10.5 gallons?

(1) $5.64 (2) $8.64 (3) $10.05 (4) $19.53 (5) $25.90

a) Estimate: **b) Answer choice:** **c) Exact solution:**

P

4. A total of 403 students participated in the sports program last year. What percent of the student body participated if there are 2,015 students at the school?

(1) 5 (2) 15 (3) 20 (4) 81 (5) 160

a) Estimate: **b) Answer choice:** **c) Exact solution:**

Answers begin on page 191.

ESTIMATING WITH FRACTIONS

The cinder blocks that Gayle was moving weighed $3\frac{1}{4}$ pounds each. If the pallet he used could hold no more than $48\frac{3}{4}$ pounds, what was the maximum number of blocks Gayle could place on it?

(1) 158 **(2)** 45 **(3)** 15 **(4)** 13 **(5)** 3

Imagine this problem without fractions. How would you solve it?

Fractions can often be difficult to work with. Sometimes just looking at fractions can confuse you! Using estimation with fractions can often help you see how to solve a problem.

Here are some steps to help you estimate with fractions:

1. Round off all numbers with fractions.

- If the fraction is $\frac{1}{2}$ or greater than $\frac{1}{2}$, drop the fraction and add 1. (round up) ... $48\frac{3}{4} \approx 49$ ($\frac{3}{4}$ is greater than $\frac{1}{2}$)
- If the fraction is smaller than $\frac{1}{2}$ or equal to $\frac{1}{2}$, just drop it. (round down) ... $3\frac{1}{4} \approx 3$ ($\frac{1}{4}$ is less than $\frac{1}{2}$)

2. Solve the problem using the estimated numbers. ... $49 \div 3 = 16\frac{1}{3}$ OR round down to 48
$48 \div 3 = 16$ (because 3 divides evenly into 48)

3. Find the answer choice closest to your estimate. ... Choice **(3) 15** is close.

4. You can check your estimate with the original numbers. ... $48\frac{3}{4} \div 3\frac{1}{4} = 15$
You estimated the correct answer.

Using rounded numbers instead of fractions can help you to "see" more clearly how to solve the problem.

Exercise 3

Directions: For each problem below, first estimate an answer using the steps outlined above. Use your estimate to choose from the list of answers. Then find the exact answer and compare it to your estimate.

1. One week Flora worked $24\frac{1}{2}$ hours at a rate of $9.70 per hour. What were Flora's earnings that week?

(1) $2,376.50 **(2)** $237.65 **(3)** $200.65 **(4)** $24.25 **(5)** not enough information is given

a) Estimate: **b) Answer choice:** **c) Exact solution:**

2. What is the difference in length (in inches) between the two rods shown here?

$24\frac{5}{8}''$

$10\frac{1}{4}''$

(1) 35 **(2)** $34\frac{1}{8}$ **(3)** $14\frac{3}{8}$ **(4)** 12 **(5)** not enough information is given

a) Estimate: **b) Answer choice:** **c) Exact solution:**

3. Josie needs to figure out how many teaspoons of sugar are in the muffins she just ate. If she ate $3\frac{1}{2}$ muffins, and each muffin contains $2\frac{1}{4}$ teaspoons of sugar, how many teaspoons of sugar did she eat in all?

(1) 2 **(2)** 4 **(3)** $7\frac{7}{8}$ **(4)** 10 **(5)** $12\frac{1}{2}$

a) Estimate: **b) Answer choice:** **c) Exact solution:**

4. A dressmaker had fabric remnants measuring $2\frac{1}{5}$ yards, $3\frac{1}{4}$ yards, $3\frac{3}{10}$ yards, and $4\frac{7}{10}$ yards. How much fabric did she have in all?

(1) $13\frac{9}{20}$ **(2)** $10\frac{3}{10}$ **(3)** $8\frac{1}{4}$ **(4)** $6\frac{1}{5}$ **(5)** $5\frac{1}{2}$

a) Estimate: **b) Answer choice:** **c) Exact solution:**

Answers begin on page 191.

WHEN AN ESTIMATE IS THE ANSWER

The Lanier family spends $122 per week on groceries, while the Lake family spends $39. About how many times more do the Laniers' groceries cost than the Lakes'?

(1) $\frac{1}{2}$ (2) 2 (3) 3 (4) 4 (5) 5

How do you know when to estimate the answer to a problem?

Do you need to compute an exact answer for the problem above?

TIP

Look for clue words such as *about* and *approximately*. When you see these words in a word problem, think about estimating the answers.

Word problems like the one above do not require you to find an exact answer. The clue that you need only an *estimated* answer for the problem above is the word *about*. In other problems you may see the word *approximately*. These clue words tell you to round off the numbers in the problem to get an *approximate* answer.

Round off the numbers above:

$122 \approx 120 \quad 39 \approx 40$

By looking quickly at these rounded numbers, you can see that 120 is *3 times* 40.

$120 \div 40 = 3$

The answer to this problem is **(3) 3** *even though* 3×39 is not *exactly* 122. Remember that the question asked "*about* how many times?"

Exercise 4

Directions: Solve the following word problems. Remember to watch out for clue words.

WN

1. The first week on her new job as a salesperson, Sandy made 4 sales. By the last week of the month she made 23 sales. About how many times did her sales increase?

(1) 4 (2) 6 (3) 19 (4) 23 (5) not enough information is given

2. It takes Earl about 7 minutes to run around the school yard. About how many times will he need to run around the yard to complete his 1 hour of physical education?

(1) 1 (2) 7 (3) 8 (4) 60 (5) not enough information is given

Questions 3–5 refer to the graph shown below.

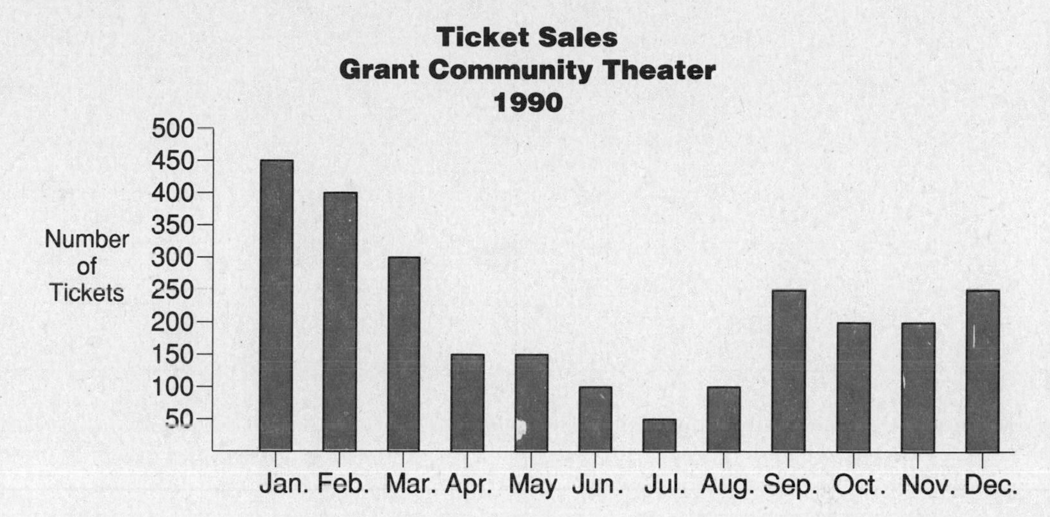

3. About how many times as many tickets were sold in January as in July?

(1) 9 (2) 10 (3) 50 (4) 450 (5) not enough information is given

4. Tickets cost $3.75 each. About how much money did the theater take in during the first 3 months of the year?

(1) $300 (2) $400 (3) $450 (4) $1,150 (5) $4,600

5. Approximately what percent of the April–May–June sales took place in June?

(1) 1% (2) 5% (3) 10% (4) 25% (5) 50%

Answers begin on page 192.

WRITING ANSWERS IN SET-UP FORMAT

To be fair, Mr. Méndez collects all customer tips and takes $10 out for each of his 2 busboys. He then divides the remainder evenly among his 3 waitresses. If he collected $134 in tips, which expression shows the amount received by each waitress?

(1) $\frac{134 - 2}{3}$

(2) $\frac{134 - 10}{3}$

(3) $(134 \div 3) - 10$

(4) $(134 \div 3) - 20$

(5) $\frac{134 - 20}{3}$

Do you remember working with "set-up" problems in Chapter 3?

How do you go about finding the right answer?

In Chapter 3 you learned about set-up questions and how they are different from other word problems. In this lesson you will again look at setups and see different ways to find a solution.

To solve the problem above, follow the steps you learned on page 36 in Chapter 3:

1. Make a statement that tells what needs to be done.	I need to subtract the busboys' pay from the total tips, then divide by the number of waitresses.
2. Fit in the numbers, just as you would write a plan for a number sentence.	I need to take two ten-dollar amounts from $134, then divide by 3.
3. Write an expression or a number sentence.	$\frac{134 - (2 \times 10)}{3}$
4. Find the answer choice that matches your set-up.	?????????????

At first you may think that no answer matches the one we came up with. The expression $\frac{134 - (2 \times 10)}{3}$ is not listed as one of the choices. However, look again:

$\frac{134 - (2 \times 10)}{3}$ can also be written as $\frac{134 - 20}{3}$—answer choice **(5)**.

TIP

If you are having trouble recognizing answers in set-up problems, decide whether your set-up can be written another way. Remember, there is often more than one way to set up a problem.

DID YOU REMEMBER THAT. . . ?

- Parentheses ( ) mean *do me first.*
- A fraction bar (—) or slash bar (/) means *divided by* (÷).
- A number written outside parentheses means multiplication: 35 (230) = 35 × 230
- Writing set-up expressions is a lot like writing equations, as you did in Chapter 5.

TIP

Practice writing set-up expressions as you solve regular word problems. This will help you work with the set-up format, **and** it will help you solve the problems.

Exercise 5

Part One

Directions: For each problem below, write **two** different set-up expressions. To indicate multiplication, you may use either × or ( ); to indicate division, use either ÷ or /.

Example: Agatha earns $5.75 per hour cleaning offices. She worked 8 hours on Monday, 8 hours on Thursday, and 10 hours on Saturday. Write an expression that shows Agatha's pay for the 3 days.

a) $5.75 (8+8+10)

b) 8($5.75) + 8($5.75) + 10($5.75)

1. Write two expressions for the distance Denny traveled if he drove for 3 hours at 55 miles per hour and another 3 hours at 60 mph.

D

2. At a garage sale, Val spent a total of $32.80. One of her purchases was a pair of candlesticks that cost $3.00 per candlestick. She spent the rest of the money on a blanket. Write two expressions to find the cost of the blanket.

3. At Cecil's Cleaning Service, every employee is expected to work 36 hours per week. Over the weekend, Byron worked $\frac{1}{4}$ of his expected hours. Write two expressions that show how many more hours Byron needs to work this week.

GE

4. Write two expressions that show the distance around the backyard shown here.

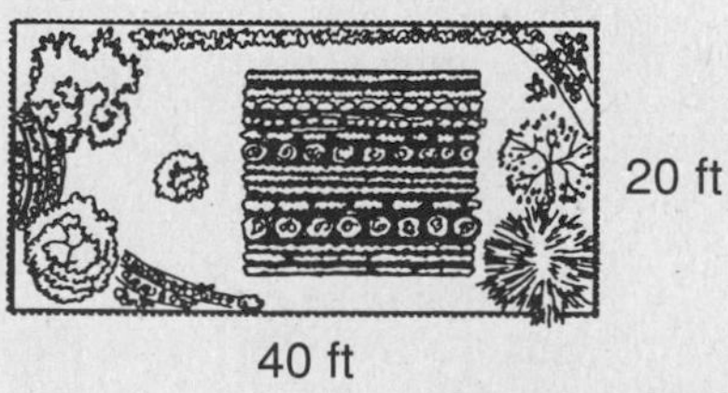

Answers begin on page 192.

COMPARING AND ORDERING NUMBERS

A survey showed that $\frac{2}{5}$ of the workers in Company A supported the union. In Company B, 55% supported it, and 75% of Company C supported it. Which of the following sequences shows the order of companies from *greatest* support to *least* support?

1) A, B, C
2) C, B, A
3) B, C, A
(4) B, A, C
(5) C, A, B

What do you need to do to solve this problem?

How can you compare fractions and percents?

Many word problems and real-life situations require you to compare numbers. This is easier when you are comparing 35 truckloads with 1,000 truckloads or 2 hours with 20 hours. But what about when the **labels** are not the same?

Here are some steps you can take when you are asked to compare and order numbers:

1. Choose *one* label for all the numbers you must compare. Try to choose the label that will make the least work for you. • • In the problem above, you could change all numbers into fractions or percents. Since 2 of the numbers are already percents, choose percent.

2. Carefully change all numbers so that they have the same label. • • Company A: $\frac{2}{5}$ = 40%
Company B: 55%
Company C: 75%

3. Now put the numbers in the correct order. • • • This problem asks for *largest* to *smallest*: 75%, 55%, 40%

4. Match the letters with the numbers • • • 75% — C
55% — B
40% — A

Answer: **(2) C, B, A**

Now try another type of comparison problem.

Based on the salaries listed below, Mathias is deciding which job to take.

A. $12,000 per year, with a 10% raise for the 2nd year
B. $6 per hour, 40 hours per week, with no raise
C. $5 per hour, 40 hours per week, with a $1,000 bonus
D. $13,000 per year, with a 5 percent raise for the second year

Which job will give Mathias the most money in his second year of employment?

(1) Job A
(2) Job B
(3) Job C
(4) Job D
(5) not enough information is given

When you do a comparing or an ordering problem, **always look at the labels first.** Make the necessary conversions; then compare.

1. Decide how to convert numbers so that they can be compared. • • • You need to convert numbers to **dollars** for the second year.

2. Carefully perform all computations. • • •

JOB A
10% of $12,000 = $1,200
$12,000 + $1,200 = **$13,200**

JOB B
$6 × 40 hours = $240 per week
$240 × 52 weeks = **$12,480**

JOB C
$5 × 40 hours = $200 per week
$200 × 52 weeks = $10,400
$10,400 + $1,000 bonus = **$11,400**

JOB D
5% of $13,000 = $650
$13,000 + $650 = **$13,650**

3. Choose the correct answer. • • • • • • • **(4) Job D**
because $13,650 is larger than all of the other second-year salaries

Exercise 6

Directions: First decide what label you will use to convert the values in each problem below. Then do the necessary conversions and solve the problem.

P

1. Tony will buy whatever automobile costs him the least. Based on the information below, which car will Tony buy?

A. a working 1984 Chevrolet for $4,800 cash
B. a new Hyundai for $6,000, with a $500 rebate
C. a 1988 Ford for $1,000 down and $100 per month for 3 years
D. a 1985 Ford for $3,000, with an added $1,500 for repairs
E. a 1989 Toyota for $6,000, less a 10% dealer discount

a) LABEL TO CONVERT ALL VALUES TO: ______________

b) ANSWER: ______________

(1) A (2) B (3) C (4) D (5) E

P

2. A survey of 5 southwestern towns showed the following results regarding the question of whether or not a nuclear power plant should be built in the area.

Town A: 62% of residents in favor
Town B: $\frac{2}{3}$ of residents in favor
Town C: 46% of residents opposed
Town D: 48% of residents opposed
Town E: $\frac{1}{3}$ of residents in favor

Which of the following lists the towns in order of *most in favor* to *least in favor*?

a) LABEL TO CONVERT ALL VALUES TO: ______________

b) ANSWER: ______________

(1) B, A, D, C, E
(2) A, C, D, B, E
(3) E, C, D, A, B
(4) C, E, A, D, B
(5) B, D, C, E, A

3. A carpenter has the following lengths of boards. To incur the least amount of waste he wants to use the boards in order from *longest* to *shortest*. Which of the following lists the correct order?

Board A: $2\frac{1}{2}$ feet
Board B: 1 yard
Board C: 40 inches
Board D: $1\frac{1}{2}$ yards
Board E: $3\frac{1}{2}$ feet

a) LABEL TO CONVERT ALL VALUES TO: ____________________

b) ANSWER: _________________________

(1) E, D, C, A, B
(2) D, E, C, B, A
(3) C, D, B, A, D
(4) B, A, D, C, E
(5) A, B, D, C, E

4. A salad preparer had the following measures of cooking oil in the kitchen pantry. Which of the following lists their order of amount from *least* to *greatest*?

1 pint = 2 cups

1 cup = 8 ounces

A. $1\frac{1}{2}$ cups of Wesson
B. $\frac{1}{2}$ of an 8-ounce bottle of Puritan
C. 1 pint of Ready-Pour
D. 3 cups of a generic brand
E. 6 ounces of Home's Cook

a) LABEL TO CONVERT ALL VALUES TO: ____________________

b) ANSWER: _________________________

(1) A, D, C, B, E
(2) D, C, B, E, A
(3) E, D, A, C, B
(4) B, E, A, C, D
(5) C, E, B, D, A

Answers begin on page 192.

WHAT TO DO WITH REMAINDERS

Problem 1:
At a construction site, the crew dug up 13,500 cubic yards of sand and soil. If a truck can haul 200 cubic yards at a time, how many trips will the crew have to take to clear the site?

What operation do you use to solve this problem?

What is the answer?

TIP

Whenever you find a remainder in your computation, READ THE QUESTION AGAIN. The question will give you a clue whether to round the remainder up or down.

You would need to divide to find the answer to this problem.

$$\begin{array}{r} 67 \text{ R } 100 \\ 200\overline{)\ 13{,}500} \\ \underline{12\ 00} \\ 1\ 500 \\ \underline{1\ 400} \\ 100 \end{array}$$

But what is the answer to the problem? 67? 67 remainder 100? $67\frac{1}{2}$?

Notice that the question above asks for *the number of trips* necessary to remove *all* the soil. After 67 trips, there are still 100 cubic yards left. The crew will need to take another trip.

The answer to the problem is **68 trips**. You needed to *round up*.

Let's look at another remainder problem.

Problem 2:
A meat-packing company uses only 40-pound crates to ship its prime cuts. If a packer has 4,300 pounds of prime beef, how many *full* crates can he ship?

(1) 10 **(2)** 100 **(3)** 107 **(4)** $107\frac{1}{2}$ **(5)** 108

?

What should you do with the remainder in this problem?

How do you know?

Step 1. Divide.	$\begin{array}{r} 107 \\ 40\overline{)4,300} \\ \underline{40} \\ 300 \\ \underline{280} \\ 20 \end{array}$
Step 2. Reread the question.	"How many *full* crates?"
Step 3. Decide what to do with the remainder.	Since the 20 pounds left over cannot make a *full* crate, do not use the remainder. You *round down.*
Step 4. Choose your answer.	The packer can ship 107 crates. **(3) 107**

What to Do with Remainders

Round Up

- When you **need to use the** remainder

Round Down

- When you **don't need to use the remainder**

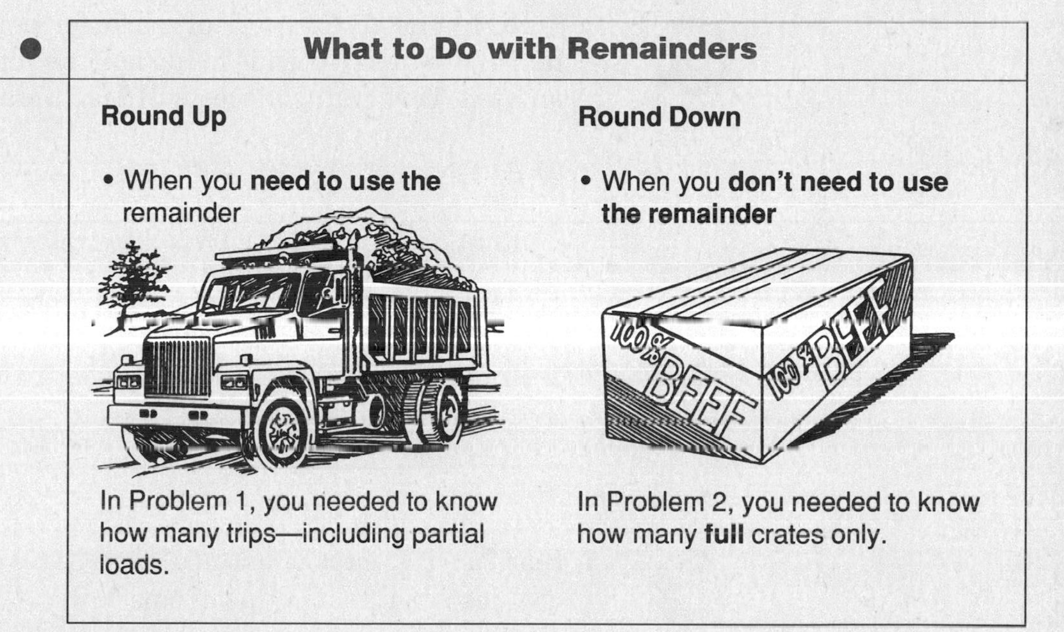

In Problem 1, you needed to know how many trips—including partial loads.

In Problem 2, you needed to know how many **full** crates only.

Exercise 7

Directions: Solve the following word problems. Pay special attention to remainders.

Example: A land development company owns 1,560 acres of land in Fairfield County. It plans to sell the land in 7-acre plots. How many of these 7-acre plots can be sold?

a) Answer with remainder: 222R6

b) Round UP or (DOWN) (circle one)

c) Why? The question asks for the number of 7-acre plots. The 6 acres are left over.

d) Answer: 222

1. The Carson Bus Lines has agreed to transport a group of schoolchildren to and from the zoo. The school is sending 6 classrooms of 30 students per class. If each bus holds 40 students, how many buses will be needed?

   a) Answer with remainder: __________

   b) Round UP or DOWN (circle one)

   c) Why? ____

   ____

   d) Answer: __________

WN

2. Mathilda needs 16 yards of fabric for her bridesmaids' dresses. The fabric she has chosen is being sold as remnants cut into 3-yard lengths. How many pieces will Mathilda need to buy?

   a) Answer with remainder: __________

   b) Round UP or DOWN (circle one)

   c) Why? ____

   ____

   d) Answer: __________

F

3. Bert has $10\frac{1}{2}$ cups of oatmeal to use in a cookie recipe. If 1 batch of cookies requires $2\frac{1}{2}$ cups of oatmeal, how many complete batches can Bert make?

   a) Answer with remainder: __________

   b) Round UP or DOWN (circle one)

   c) Why? ____

   ____

   d) Answer: __________

Answers begin on page 192.

MIXED REVIEW

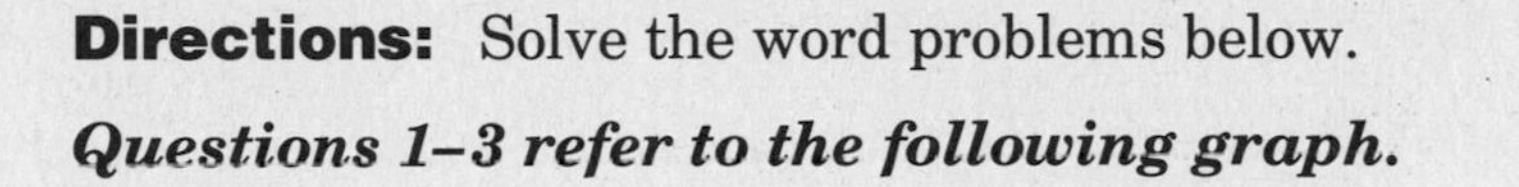

Directions: Solve the word problems below.

Questions 1–3 refer to the following graph.

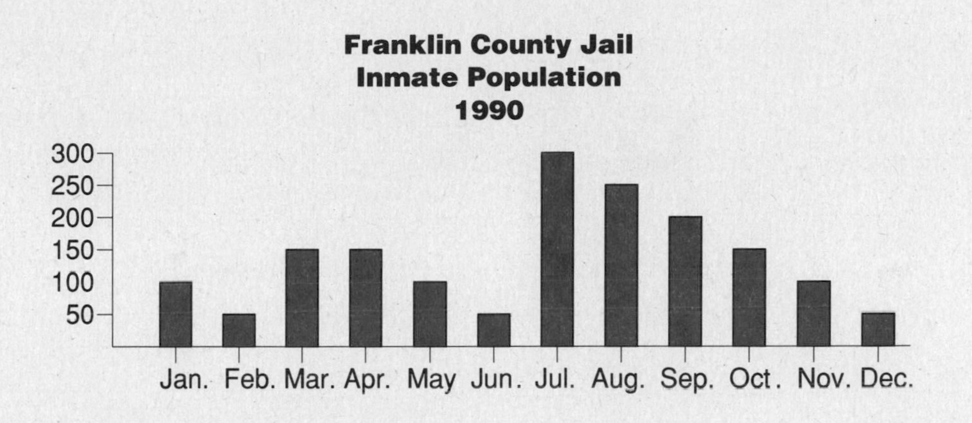

1. What was the average number of inmates at Franklin County during the first half of the year?

(1) 600 (2) 300 (3) 200 (4) 100 (5) 50

WN

2. In February 1990, how many Franklin County inmates were charged with drug offenses?

(1) 5 (2) 10 (3) 25 (4) 50 (5) not enough information is given

3. A total of 69% of the Franklin County inmates in July were repeat offenders. Which of the following expressions shows the the number of repeat offenders in July?

(1) .69 (100)
(2) 300 – .69 (300)
(3) .69 ÷ 300
(4) .69 + 300
(5) .69 (300)

4. Your car's gas tank can hold 14 gallons of gas. The gas gauge measures $\frac{1}{2}$ full. About what will it cost to fill up the tank if gas costs $1.17 per gallon?

(1) $.80
(2) $8
(3) $16
(4) $32
(5) not enough information is given

F

5. Rafael has $6\frac{1}{2}$ more hours to work today. He worked 9 hours yesterday and 10 hours the day before. What more do you need to know to find the total hours Rafael worked in the 3 days?

(1) his hours per week
(2) his wages per hour
(3) the hours he has already worked today
(4) his average daily hours
(5) the time he got to work today

P

6. Of the 4,450 residents of Honeysuckle Hill, 40% say they will vote for Matthew Locke for mayor and 30% say they will vote for Gary Meyers. The rest of the residents are undecided. Which of the following expressions shows the number of undecided residents?

(1) 4,450 + .70 (4,450)
(2) 4,450 – .70 (4,450)
(3) 4,450 – .1 (4,450)
(4) 4,450 + .4 (4,450)
(5) 4,450 – .4 (4,450)

7. Pantsuit Heaven took 30% off everything in stock last weekend. Which of the following expressions shows the amount of money Naomi paid for a pantsuit that was originally priced at $59.99?

(1) $59.99 (.30)
(2) $59.99 + .30 ($59.99)
(3) $59.99 – .30
(4) $59.99 – .30 ($59.99)
(5) $59.99/30

P

8. Bill wants to buy the least expensive cereal he can find. Based on the information below, which should he buy?

Cereal A: $1.98, minus a 30-cent coupon
Cereal B: $2.09, minus a 35-cent coupon
Cereal C: $2.20, minus a 10% store discount
Cereal D: $1.69
Cereal E: $2.25, including a 35-cent discount on a box of doughnuts

(1) A (2) B (3) C (4) D (5) E

P

9. A new can of roach killer is advertised as containing "25 percent more spray" for the regular price of $3.39. If the previous can contained 28 ounces of spray, how many ounces does the new can hold?

(1) 4 (2) 4.24 (3) 7 (4) 35 (5) 53

M

10. How many 6-inch-long pieces of wood are needed to cover 1 yard of garden border?

(1) $\frac{1}{2}$ (2) 5 (3) 6 (4) 7 (5) not enough information is given

Answers begin on page 193.

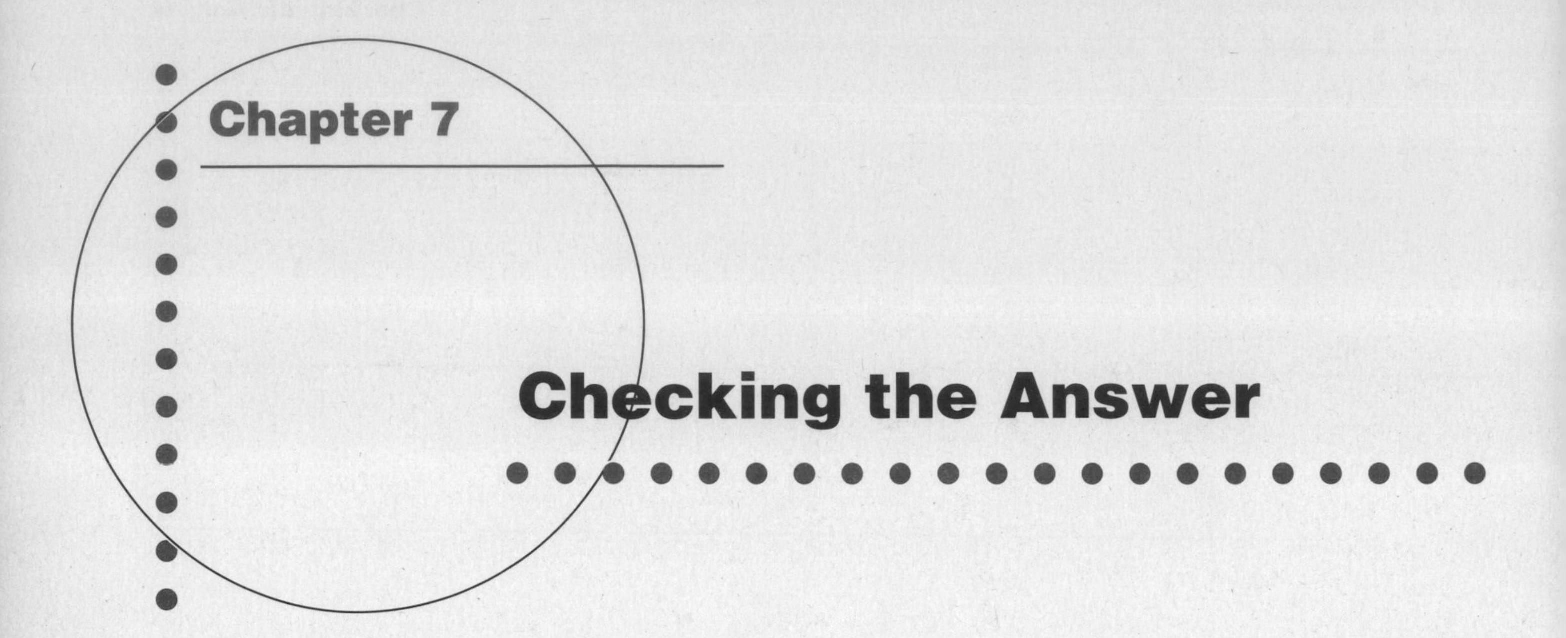

Chapter 7

Checking the Answer

DID YOU ANSWER THE QUESTION?

By 2:00 P.M., each worker at Smart's Dry Cleaners had pressed 120 shirts. To finish out the day, the 3 workers evenly divided the remaining 60 shirts. By the end of the day, how many shirts did each worker press?

(1) 180 **(2)** 140 **(3)** 60 **(4)** 20 **(5)** not enough information is given

How many operations does it take to solve this problem?

Do you think 20 is the correct answer? or 140?

If you found the answer of this problem to be 20, you have made a common mistake. You will see your error and help avoid similar mistakes in the future by checking to see if you *answered the question asked*.

TIP

Rereading the question is very important in multi-step problems like the one here. The solution for the first step is often listed as an answer choice for the problem, so watch out!

1. First find how many shirts each worker pressed after 2:00 P.M.	$60 \div 3 = 20$ shirts
2. But don't stop there! Reread the question.	The question asks for the total that each worker pressed—not the amount after 2:00 P.M.
3. Now add the two amounts.	120 shirts *(by 2:00 P.M.)* + 20 shirts *(after 2:00 P.M.)* 140 shirts per worker

Answer the Question Asked

Let's take a look at another problem. Pay special attention to what you are being asked to find.

Jacqueline bought a leather jacket during a $\frac{1}{3}$-off sale at a department store. The jacket was originally priced at \$90. How much did Jacqueline save by buying the jacket on sale?

(1) \$120 **(2)** \$90 **(3)** \$60 **(4)** \$30 **(5)** \$3

What is the answer to this problem?

Does your solution answer the question asked?

1. To get the answer, find $\frac{1}{3}$ of 90.	$\$90 \div 3 = \$30$
2. If you think you should subtract \$30 (amount saved) from \$90 (original price), REREAD THE QUESTION.	The question asks for the *amount saved*—NOT the amount paid! Answer: **(4) \$30**

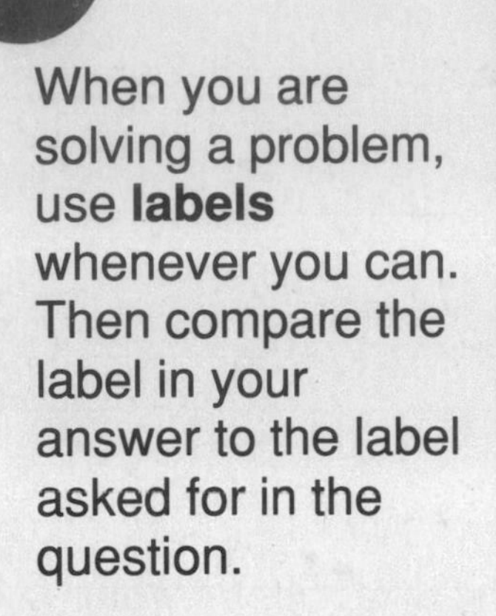

TIP

When you are solving a problem, use **labels** whenever you can. Then compare the label in your answer to the label asked for in the question.

Exercise 1

Directions: First solve each problem, paying close attention to what you are being asked to find. Then look at the *incorrect* answer given below "your solution" and write down *what question* this incorrect solution is answering. This will help you focus on answering the question that was asked.

Example: In 1988, Caroline's neighborhood association had 130 members. Over the next two years, that number increased by 20%. How many members did the association have in 1990?

(1) 26 (2) 104 (3) 146 (4) 156 (5) not enough information is given

a) Your solution: Step 1: $130 \times .20 = 26$ members
Step 2: $130 + 26 = 156$ members
Answer: **(4) 156**

b) Choice (1) is incorrect because it answers the question: By how many members did the association increase?

1. A land-developing company owned 1,350 acres in Red County last year. This year, the company added 200 acres in March, 750 acres in June, 90 acres in September, and 110 acres in November. How many total acres were added this year?

(1) 2,500 (2) 1,150 (3) 1,060 (4) 950 (5) 200

a) Your solution:

b) Choice (1) is incorrect because it answers the question: ____________

____

2. Maureen started the week with $404.00 in the bank. During the week she wrote checks for $37.50, $18.97, and $144.00. She also deposited a tax refund of $200.00. How much did Maureen have in her account at the end of the week?

(1) $200.47 (2) $337.53 (3) $403.53 (4) $604.00 (5) not enough information is given

a) Your solution:

b) Choice (1) is incorrect because it answers the question: ____________

____

P

3. Sixty-five percent of the inmate population at Reeves Correctional Facility is white. The number of black and other minority inmates is 700. What is the total number of inmates at Reeves?

(1) 4,000 (2) 2,000 (3) 1,300 (4) 70 (5) 35

a) Your solution:

b) Choice (3) is incorrect because it answers the question: ____________

____

P

4. As an employee of a store, Floyd gets 15% off anything he purchases. If he bought a suitcase priced at $35 and a pair of shoes priced at $40, what did Floyd actually pay for these 2 items (before taxes)?

(1) $75 (2) $63.75 (3) $28.75 (4) $23.75 (5) $11.25

a) Your solution:

b) Choice (5) is incorrect because it answers the question: ____________

____

5. To a bucket that contained 1 quart of water, a janitor added $\frac{3}{4}$ quart of ammonia and 2 cups of liquid detergent. How many *quarts* of liquid were in the bucket?

(1) $2\frac{1}{4}$ (2) $3\frac{3}{4}$ (3) 5 (4) 10 (5) not enough information is given

a) Your solution:

b) Choice (3) is incorrect because: ____________

____

Answers begin on page 193.

IS THE ANSWER REASONABLE?

Marcie's checking account balance was $890. She then wrote checks for $110 to the electric company, $42.33 to the telephone company, and $21.90 to a department store. What is her new balance?

(1) $1,605.77 **(2)** $1,064.23 **(3)** $954.23 **(4)** $715.77 **(5)** $315.77

Suppose you added $110, $42.33, and $21.90 to $890 and chose (2) $1,064.23 as your answer. How can you tell that you have done something wrong?

One part of checking your answer is deciding whether your solution is **reasonable**—that is, does it make sense?

Why is answer choice **(2)** unreasonable? If Marcie started with $890 and wrote checks (subtracted), it doesn't make sense that her new balance is **larger** than the opening balance!

Here is the correct solution:

$890.00	beginning balance
– $110.00	check to electric company
$780.00	
– $42.33	check to phone company
$737.67	
– $21.90	check to department store
$715.77	

Answer: **(4) $715.77**

A quick check will tell you that $715.77 is a reasonable amount to have left in the account. You can check by rounding the numbers.

	$900	– $100	– $40	– $20 = $740 estimated
rounded	↑	↑	↑	↑
from	890	110	42.33	21.90

$740 ≈ choice **(4) $715.77**

Now take a look at another problem on the top of page 133:

TIP

Once you have solved a problem, make a statement out of the question by inserting your answer. Think about your statement. Does it make sense?

In the example to the right, you could see that this sentence does **not** make sense:

"James, who was 34 inches at age 2, is now 17 inches tall."

Some people believe that if you double a child's height at age two you will have the height of that child as an adult. According to this rule, if James was 34 inches at age 2, how many inches would he be as an adult?

(1) 2 **(2)** 17 **(3)** 36 **(4)** 51 **(5)** 68

Suppose you divided 34 inches by 2 and chose (2) 17 as your answer. Is 17 inches a reasonable height for an adult?

Quickly compare your answer to the facts given in the problem. Seventeen inches is *not* a reasonable height for an adult—especially if he was 34 inches tall as a child.

The correct solution is found this way:

$$\underset{\text{age 2}}{34} \times \underset{\text{doubled}}{2} = \underset{\text{adult}}{68 \text{ inches}}$$

Does this make sense?

34 inches

17 inches

Answer: **(5) 68**

Exercise 2

Directions: First solve each problem, paying close attention to what you are being asked to find. Then look at the *incorrect* answer given below "your solution" and write down why the answer choice is unreasonable.

Example: Nancy started work at 5:00 P.M. She spent 2 hours dusting, $1\frac{1}{2}$ hours emptying wastebaskets, and 1 hour sweeping. What time did Nancy finish this work?

(1) 12:30 P.M. **(2)** 1:30 P.M. **(3)** 7:30 P.M. **(4)** 9:30 P.M. **(5)** 10:00 P.M.

a) Your solution:

Step 1: 2 hr + $1\frac{1}{2}$ hr + 1 hr = $4\frac{1}{2}$ hr

Step 2: 5:00 P.M. + $4\frac{1}{2}$ hr = 9:30 P.M.

Answer: **(4) 9:30 P.M.**

b) Choice (1) is unreasonable because: It is $4\frac{1}{2}$ hours earlier, not later, than when Nancy started.

1. Della went on a new exercise program and diet. She lost 23 pounds. Before this loss she weighed 157 pounds. What is Della's new weight, in pounds?

(1) 200 (2) 180 (3) 155 (4) 134 (5) not enough information is given

a) Your solution:

b) Choice (2) is unreasonable because: ______________________________

____

2. Last week Sharen worked the hours shown in the chart. What was the *average* number of hours she worked daily over this 7-day period?

Monday	8 hours
Tuesday	6
Wednesday	3
Thursday	8
Friday	5
Saturday	3
Sunday	2

(1) 40 (2) 35 (3) 7 (4) 5 (5) 2

a) Your solution:

b) Choice (2) is unreasonable because: ______________________________

____

WN

3. Tim drove 60 miles per hour for 4 hours and 50 miles per hour for 5 hours. How many miles did he travel in all on this trip?

(1) 9 (2) 10 (3) 15 (4) 25 (5) 490

a) Your solution:

b) Choice (4) is unreasonable because: ______________________________

____

Answers begin on page 194.

CHECKING YOUR COMPUTATION

Mrs. Ruben paid $1.41 for 3 bottles of vinegar and $5.07 for 3 pounds of ground meat. How much would she pay for 1 bottle of vinegar and 1 pound of meat?

(1) $.47 **(2)** $1.69 **(3)** $2.16 **(4)** $6.48 **(5)** $7.92

What operations do you use to find the solution?

How do you check your work?

Suppose you are sure you have answered the question being asked, and you are certain that your answer is reasonable. However, you can't find your solution listed in the choice of answers.

One final step in checking your answer is *making sure you have done the computation correctly.* Follow the examples below to see how to do this check:

Perform each operation carefully.

Step 1:

$\$1.41 \div 3 = \$.47$ per bottle

Check: $\$.47 \times 3 = \$1.41$ ✓

Step 2:

$\$5.07 \div 3 = \$1.69$ per pound

Check: $\$1.69 \times 3 = \$5.07$ ✓

Step 3:

$\$.47 + \$1.69 = \$2.16$ total

Check: $\$2.16 - \$1.69 = \$.47$ ✓

DID YOU REMEMBER THAT . . . ?

- To check an addition problem, you must subtract.
- To check a subtraction problem, you must add.
- To check a division problem, you must multiply.
- To check a multiplication problem, you must divide.

Exercise 3

Directions: All of these problems are wrong! Find the computation errors by using the checkup steps given above.

Example:

	Check:	Corrected:	Recheck:
$\begin{array}{r} 345 \\ +428 \\ \hline 763 \end{array}$	$\begin{array}{r} 763 \\ -428 \\ \hline 335 \end{array}$	$\begin{array}{r} 345 \\ +428 \\ \hline 773 \end{array}$	$\begin{array}{r} 773 \\ -428 \\ \hline 345 \end{array}$

1. $5 \overline{)975}$ = 185   Check: Corrected: Recheck:

2. $\begin{array}{r} 105 \\ \times 30 \\ \hline 31{,}500 \end{array}$ Check: Corrected: Recheck:

3. $\begin{array}{r} 1{,}009 \\ -437 \\ \hline 562 \end{array}$ Check: Corrected: Recheck:

4. $23 \overline{)11{,}201}$ = 481   Check: Corrected: Recheck:

Answers begin on page 194.

MIXED REVIEW

Directions: Solve the following problems.

WN

1. A customer put $25 down for a coat he put on layaway. He agreed to pay off the balance in 3 installments over the next 3 months. What is the total cost of the coat?

(1) $25 (2) $75 (3) $100 (4) $125 (5) not enough information is given

WN

2. A 2-foot square carton holds 24 cans of pie filling. If a packer has 640 cans to pack, how many of these cartons can he fill completely?

(1) 24 (2) 25 (3) 26 (4) 27 (5) 15,360

WN

3. According to the graph, which of the states shown had the highest per-capita personal income in 1987?

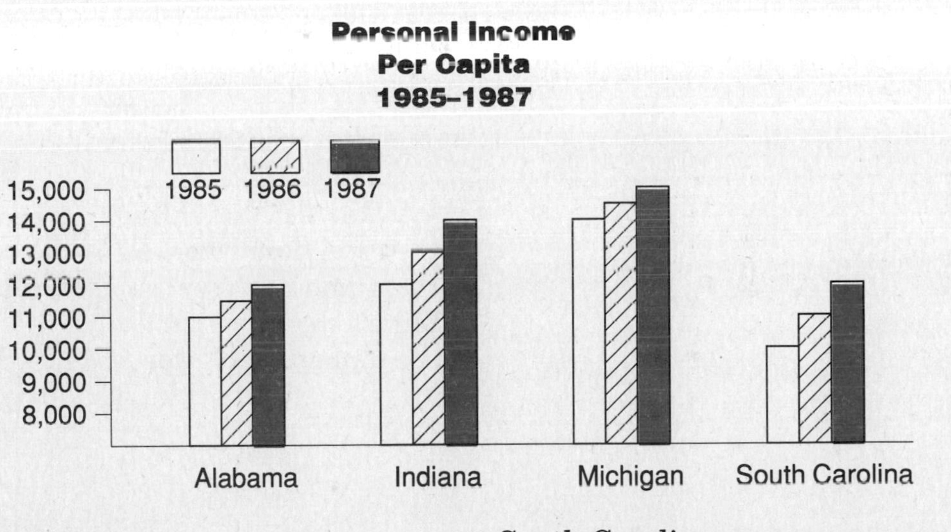

(1) Alabama
(2) Indiana
(3) Michigan
(4) South Carolina
(5) not enough information is given

WN

4. A schoolteacher bought 6 packages of a dozen pencils each and distributed them to her 18 students. Which of the following expressions shows the number of pencils received by each student?

(1) $6 \div 18$
(2) $18 \div 6$
(3) $12\,(18 \div 6)$
(4) $18(6 \times 12)$
(5) $\frac{6 \times 12}{18}$

WN

5. For every 2 job candidates that Ned turns away he usually hires 3. If, in a given month, Ned hires 12 people, how many candidates did he turn away?

(1) 1 (2) 2 (3) 4 (4) 8 (5) 24

D

6. Mr. Tyson cuts a 14-inch piece of wire into lengths of .35 inches. How many pieces of wire does he get?

(1) 4.9 (2) 40 (3) 80 (4) 400 (5) 540

F

7. McCarthy's Deli has a $\frac{1}{4}$-off sale on all prepared salads. How much does Hannah save by buying 2 pounds of potato salad originally priced at $1.50 per pound?

(1) $.37 (2) $.75 (3) $1.13 (4) $2.00 (5) $2.25

M

8. The chart shows the regular hours worked by Mike's construction crew. What more do you need to know to find out how much money Mike paid out for payroll that day?

Henry Elias	8:00 A.M.–5:00 P.M.
Yolanda Grace	7:00 A.M.–4:00 P.M.
Mitch Stewart	7:00 A.M.–4:00 P.M.
Ronald Jones	7:00 A.M.–4:00 P.M.

(1) the total hours worked by the crew
(2) the number of overtime hours put in
(3) Mike's weekly payroll
(4) the amount each worker is paid per hour
(5) Mike's expenses other than payroll

Questions 9 and 10 refer to the following information.

When she shops, Ellen buys only the most economical cleaning supplies, frozen foods, and canned goods. Then she buys 3 pounds of whatever meat is on sale.

Last week the supermarket had Luxury dish soap at $1.60 and Most dish soap at $1.92. In addition, 16 ounces of Nature frozen corn was on sale for $2.03, while Better frozen corn was priced at $2.29 per pound.

9. Before sales tax, approximately how much did Ellen pay for 1 bottle of dish soap and 1 bag of frozen corn?

(1) $8.00 (2) $4.50 (3) $3.50 (4) $43.00 (5) not enough information is given

D

10. If round steak was on sale for $2.49 per pound, how much did Ellen pay for meat this week?

(1) $2.49 (2) $4.98 (3) $7.47 (4) $9.96 (5) not enough information is given

Answers begin on page 194.

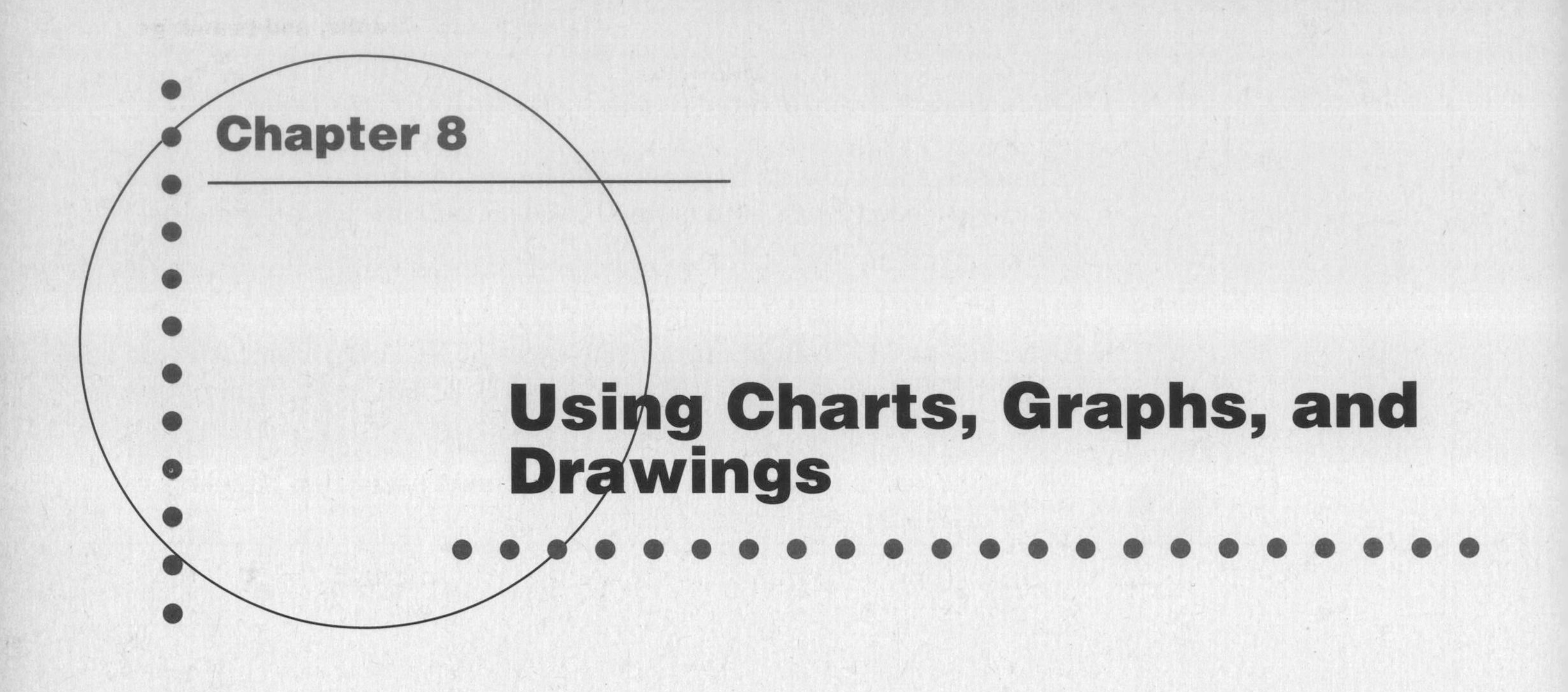

Chapter 8

Using Charts, Graphs, and Drawings

"READING" A PICTURE

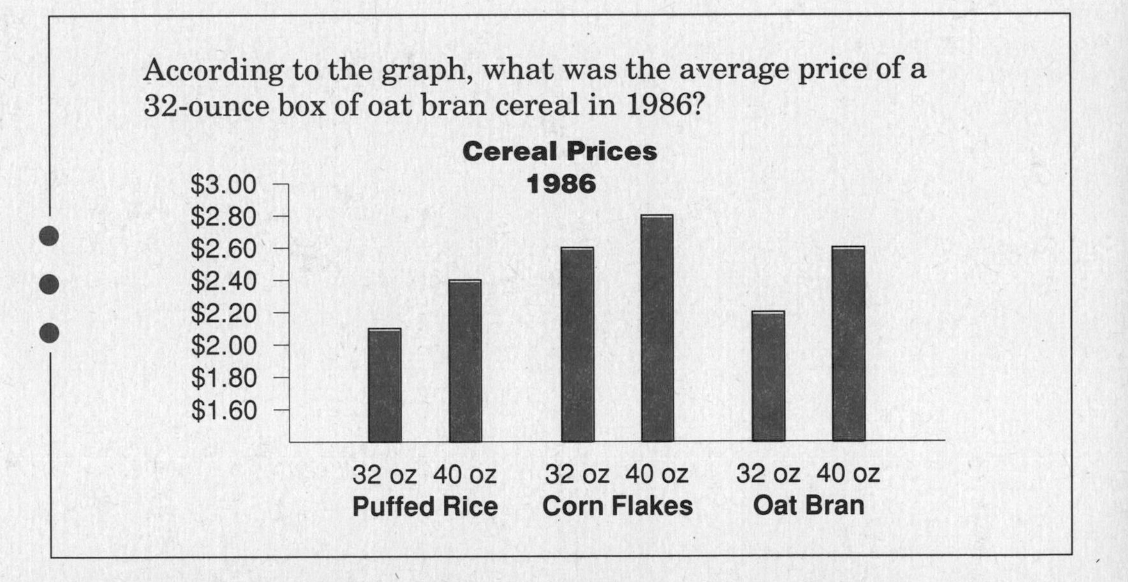
According to the graph, what was the average price of a 32-ounce box of oat bran cereal in 1986?

Could you answer the question without looking at the graph?

How do you go about finding the right information on the graph?

Graphs, charts, tables, and drawings are all ways to show a "picture" of some information. Rather than use long lists of words and numbers, people can use these tools to make information easy to understand and compare. Many word problems use graphs, charts, and pictures because they are "real-life" examples of math in action.

To answer the problem on page 140 you must

- **"read" the accompanying graph.** The problem states "according to the graph." You are being asked to use information from the drawing—not from your own experience.
- **find the year 1986.** The title to the graph indicates that **all** of the information is for 1986.
- **find the "oat bran" category of cereals listed.** Oat bran is represented by the last two bars on the graph.
- **find the correct box size in the oat bran category.** The 32-ounce size is the first bar for oat bran.

Now that you have found the spot where your information is located,

- **scan across the graph** to the vertical axis where the actual prices are listed, as shown below.

- **decide what the price is.** In this problem the 32-ounce oat bran cereal bar rises exactly to the **$2.20** mark.

Now take a look at another problem, based on the same graph.

According to the graph, what was the average price *per ounce* of puffed rice cereal in the 40-ounce package?

(1) $2.40 (2) $2.20 (3) $.20 (4) $.06 (5) $.02

How is this problem different from the one on page 140?

What are the operations needed to solve it?

TIP

Some word problems will ask you only to find a value on a graph or chart. Other word problems will require you to find a value and do some computation with this value.

To solve this problem you'll need to do all the same operations you did for the last problem. However, there is another step.

Once you have found the package price ($2.40), you need to find the unit price (price per ounce).

$$40\overline{)\$\,2.40}\quad\$\,.06 \leftarrow \text{price per ounce}$$

40 ↑ number of ounces; $ 2.40 ↑ package price

The correct answer is **(4) $.06**.

Using Charts

Take a look at a problem based on a chart of information.

People responding to a magazine questionnaire provided the information in the chart at right. If 1,000 people responded, how many of them pay under $50 per week for child care?

(1) 10 **(2)** 50 **(3)** 100 **(4)** 150 **(5)** 1,000

What People Pay for Child Care 1990

Under $50	10%
$50–$100	52%
$101–$200	33%
Over $200	5%

What are the steps involved in getting the correct solution?

Do you need to just locate information, or do you need to compute as well?

Once you have found the category for "under $50," read across and find the correct percent. The chart shows that 10% of the people paid under $50 per week.

What People Pay for Child Care 1990

Under $50	**10%**
$50–$100	52%
$101–$200	33%
Over $200	5%

10% of 1,000 =

$.10 \times 1{,}000 = 100$ (OR $\frac{1}{10} \times 1{,}000 = 100$) →

The correct answer is **(3) 100**.

Exercise 1

Directions: Solve the following problems using the information provided.

Questions 1–3 refer to the graph.

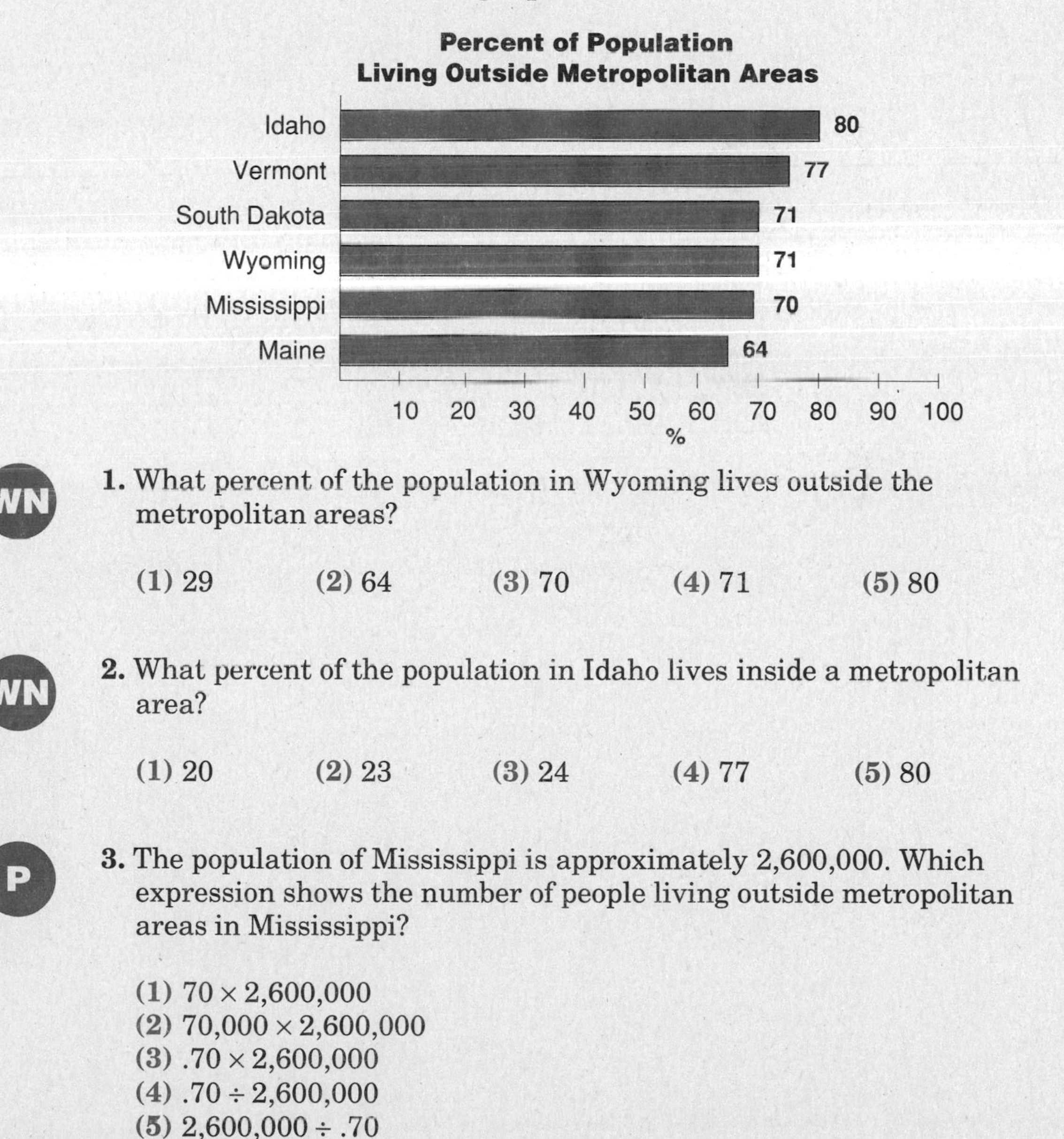

WN **1.** What percent of the population in Wyoming lives outside the metropolitan areas?

(1) 29 (2) 64 (3) 70 (4) 71 (5) 80

WN **2.** What percent of the population in Idaho lives inside a metropolitan area?

(1) 20 (2) 23 (3) 24 (4) 77 (5) 80

P **3.** The population of Mississippi is approximately 2,600,000. Which expression shows the number of people living outside metropolitan areas in Mississippi?

(1) $70 \times 2{,}600{,}000$
(2) $70{,}000 \times 2{,}600{,}000$
(3) $.70 \times 2{,}600{,}000$
(4) $.70 \div 2{,}600{,}000$
(5) $2{,}600{,}000 \div .70$

Questions 4–6 refer to the graph below.

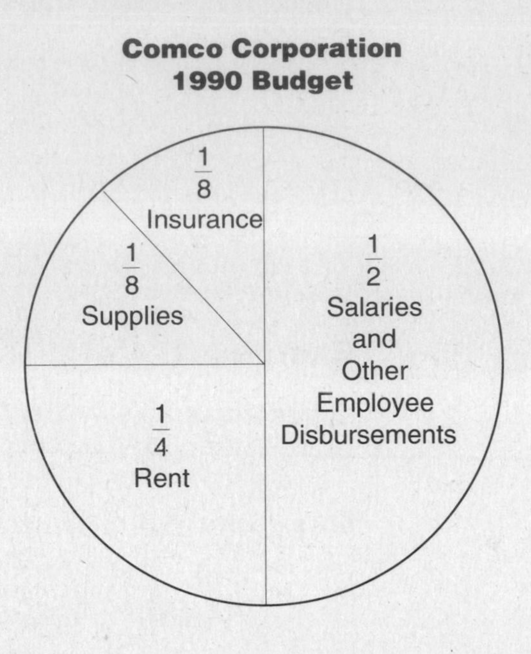

F

4. What fraction of the budget for Comco Corporation is *not* spent on rent?

(1) $\frac{1}{8}$ **(2)** $\frac{3}{8}$ **(3)** $\frac{1}{2}$ **(4)** $\frac{5}{8}$ **(5)** $\frac{3}{4}$

F

5. How many times more money does Comco spend on rent and insurance than it does on supplies?

(1) $\frac{1}{8}$ **(2)** 2 **(3)** 3 **(4)** 4 **(5)** 5

F

6. If the entire budget for Comco in 1990 was $800,000, how much money did the company spend on insurance?

(1) $\frac{1}{8}$ **(2)** $7,000 **(3)** $10,000 **(4)** $100,000 **(5)** $700,000

Answers begin on page 195.

"READING BETWEEN THE LINES" ON GRAPHS

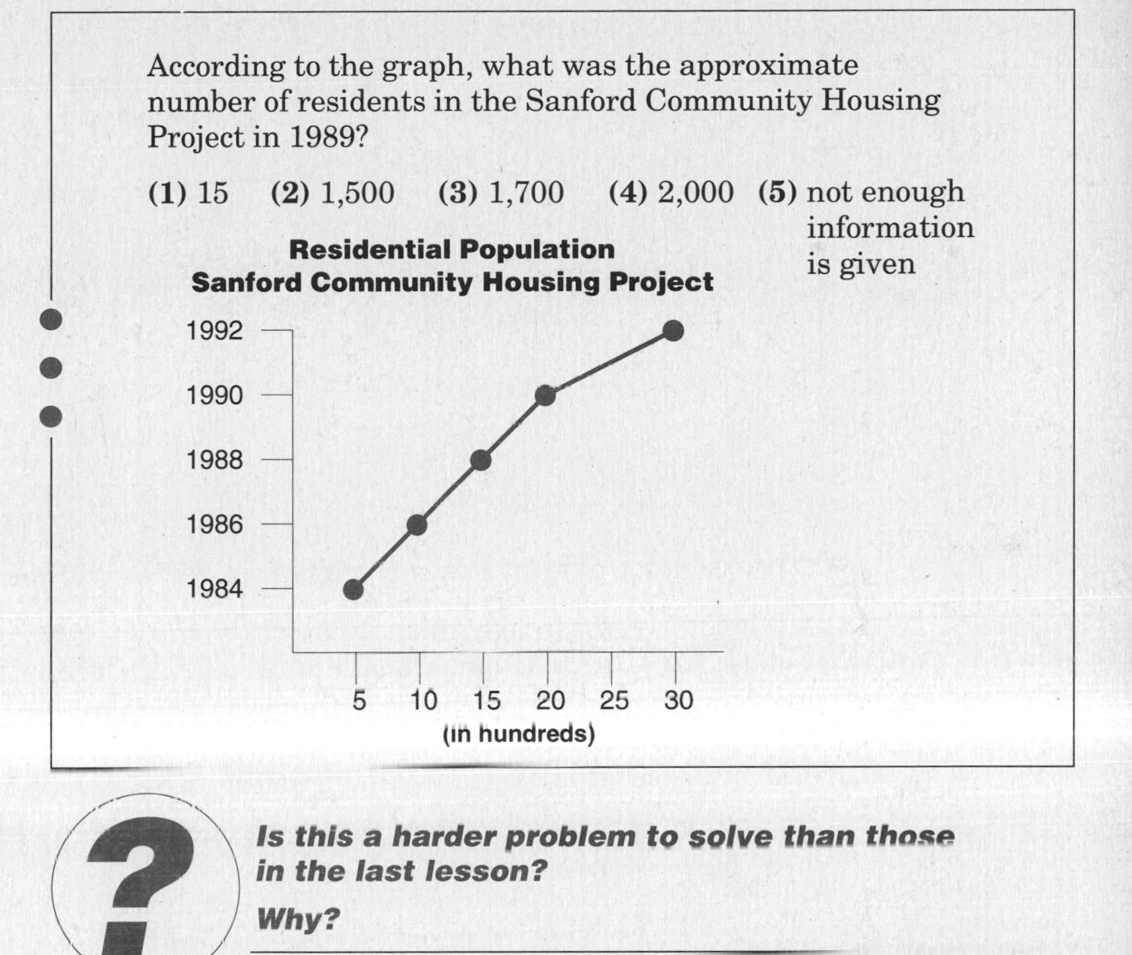

According to the graph, what was the approximate number of residents in the Sanford Community Housing Project in 1989?

(1) 15 **(2)** 1,500 **(3)** 1,700 **(4)** 2,000 **(5)** not enough information is given

Is this a harder problem to solve than those in the last lesson?

Why?

You need to find 1989 and look across to find the correct number of residents. The tricky part of this problem is that you do **not** see 1989 clearly labeled on the graph.

To solve this type of graph problem, you'll need to **estimate**:

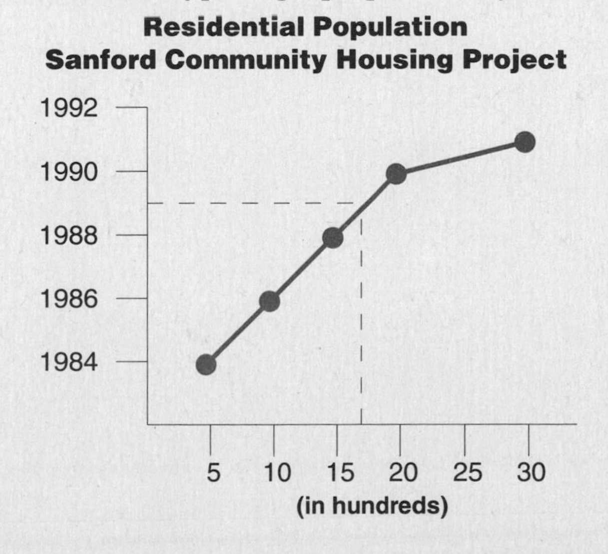

The answer is between 15 hundred (1,500) and 20 hundred (2,000). The value you are trying to find is a little closer to the 15 than the 20; therefore, a good estimate would be **(3) 1,700 residents**.

The next exercise will give you more practice in estimating with graphs.

Exercise 2

Directions: Solve the following problems using the graphs provided.

Questions 1–3 refer to the graph below.

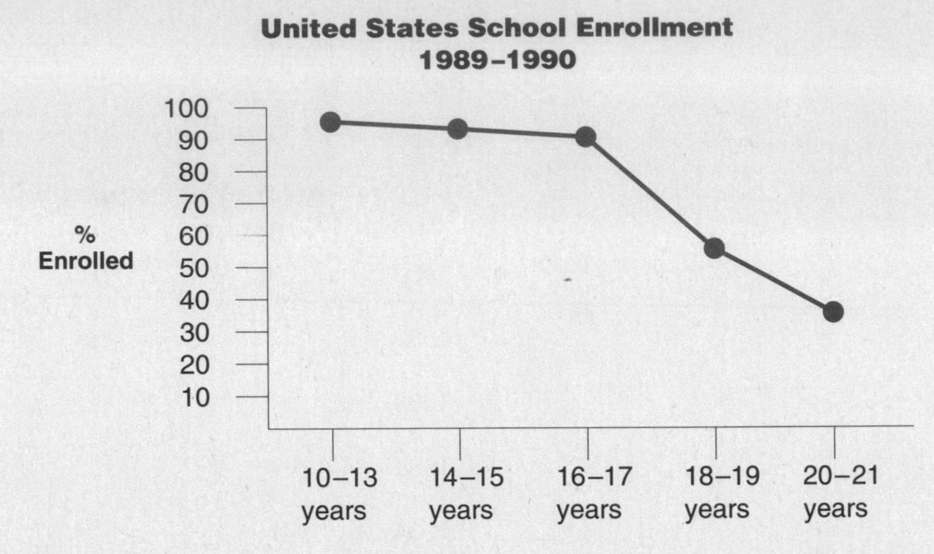

WN

1. What age group (between 10 and 21 years) had the smallest percent enrolled in school?

(1) 10–13 (2) 14–15 (3) 16–17 (4) 18–19 (5) 20–21

2. Approximately what percent of all 14- and 15-year-olds were enrolled in school in 1989?

(1) 9 (2) 31 (3) 51 (4) 91 (5) 99

P

3. If school enrollment in the town of Westville is typical of school enrollment across the United States, approximately how many of Westville's 5,500 18- and 19-year-olds are enrolled in school?

(1) 55 (2) 100 (3) 3,025 (4) 4,000 (5) 5,500

Questions 4–6 refer to the following graph.

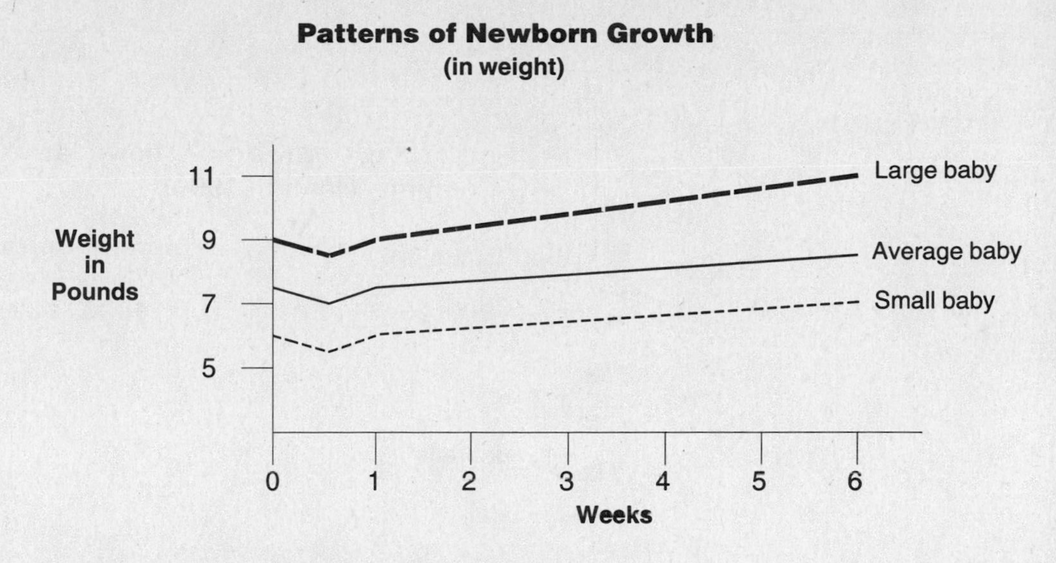

4. According to the graph, most newborns

(1) gain weight during their first week, then lose weight.
(2) lose weight during their first week, then gain weight.
(3) lose weight during their first 6 weeks.
(4) stay the same weight during their first 6 weeks.
(5) weigh less than 6 pounds at birth.

5. By the age of 6 weeks, what is the approximate difference in pounds between a large baby and a small baby?

(1) 0 (2) 4 (3) 6 (4) 8 (5) 11

6. At the end of the first week, what is the approximate difference in pounds between an average baby and a small baby?

(1) 2 (2) 4 (3) 6 (4) 8 (5) 9

Answers begin on page 195.

READING GRAPHS AND CHARTS CAREFULLY

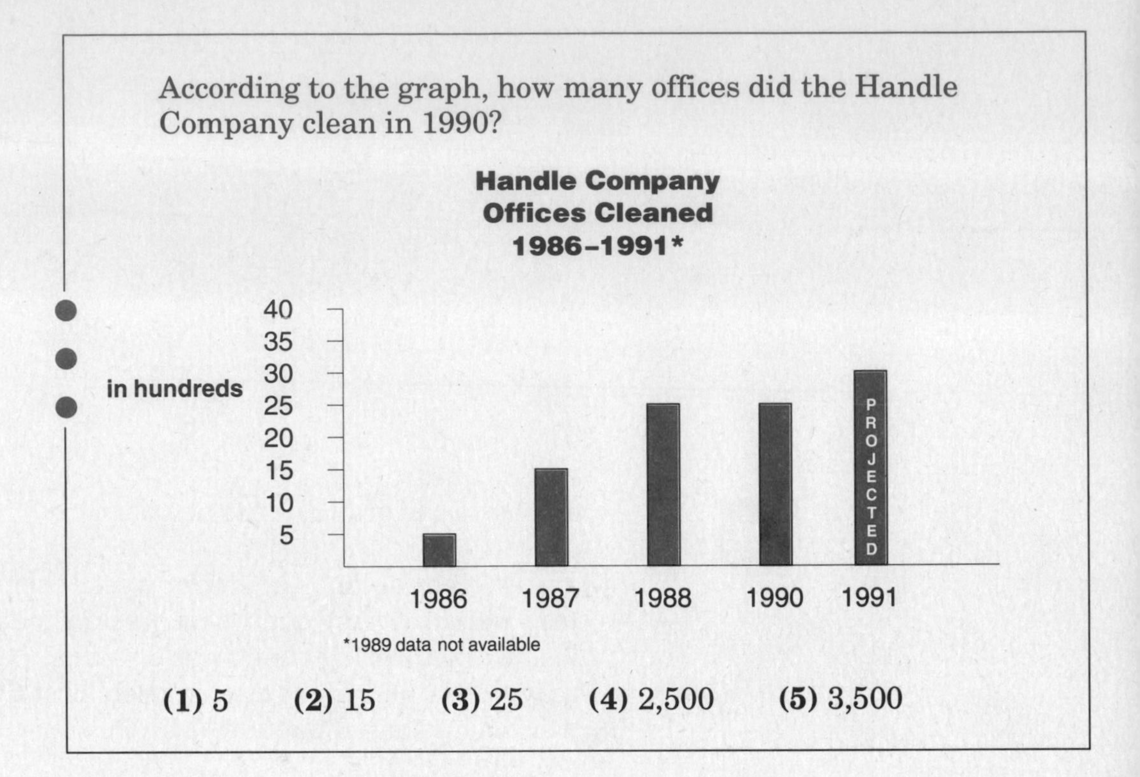

According to the graph, how many offices did the Handle Company clean in 1990?

(1) 5 **(2)** 15 **(3)** 25 **(4)** 2,500 **(5)** 3,500

John quickly found the 1990 bar on the graph and read across to the value of 25. He chose **(3) 25** as his answer.

What mistake did John make?

When you solve a problem based on a graph, you must be careful to read **all** of the information given. Although John correctly found 25 on the graph, he failed to see the label *(in hundreds)* included on the side of the graph. The correct answer to this problem is **(4) 2,500**. ($25 \times 100 = 2{,}500$.)

Try another problem.

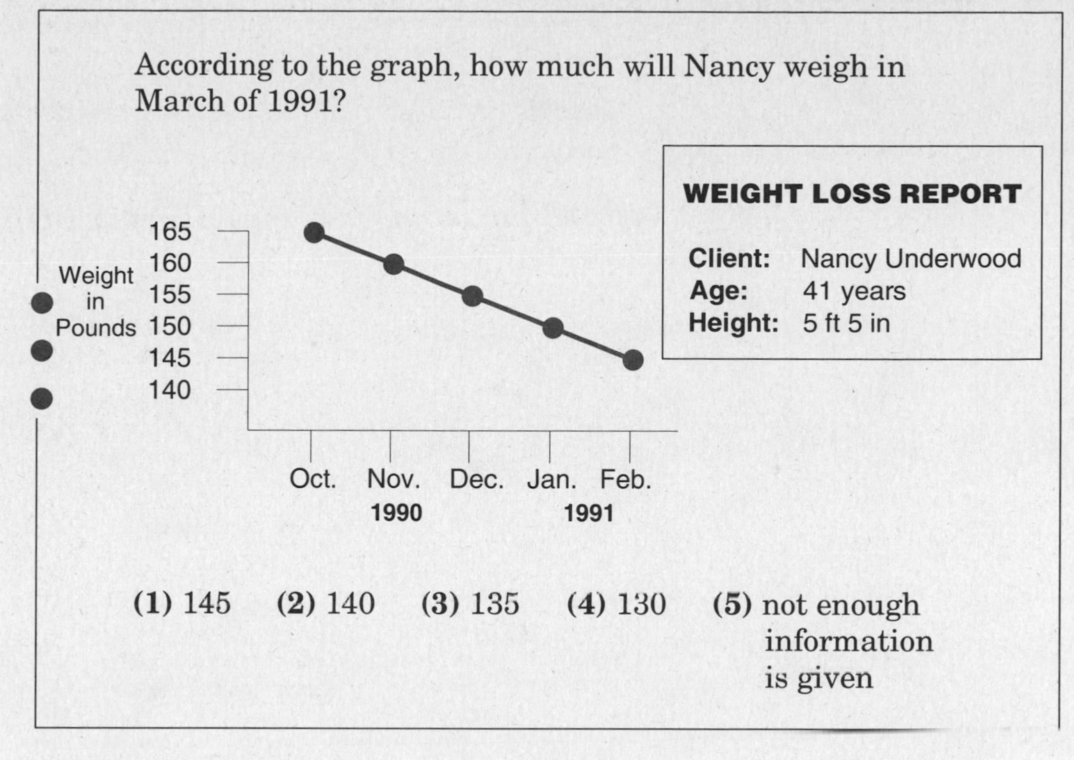

According to the graph, how much will Nancy weigh in March of 1991?

(1) 145 **(2)** 140 **(3)** 135 **(4)** 130 **(5)** not enough information is given

Pauline looked at the graph and noticed that Nancy was losing weight at a rate of 5 pounds per month. She subtracted 5 pounds from the 145 that Nancy weighed in February 1991. She chose **(2) 140** as her answer.

What mistake did Pauline make?

Pauline made one major mistake. The graph does not provide *any* information about March of 1991. She cannot assume that Nancy will weigh 5 pounds less in March than in February. (Perhaps Nancy started a different diet and lost 8 pounds instead; or perhaps 145 pounds was her desired weight.) The correct answer to this problem is **(5) not enough information is given**.

When using a graph, use *all* the labels and information provided. However, don't make assumptions or predictions unless you are specifically asked to.

Exercise 3

Directions: Use the information provided to answer the following problems.

Questions 1–3 refer to the following graph.

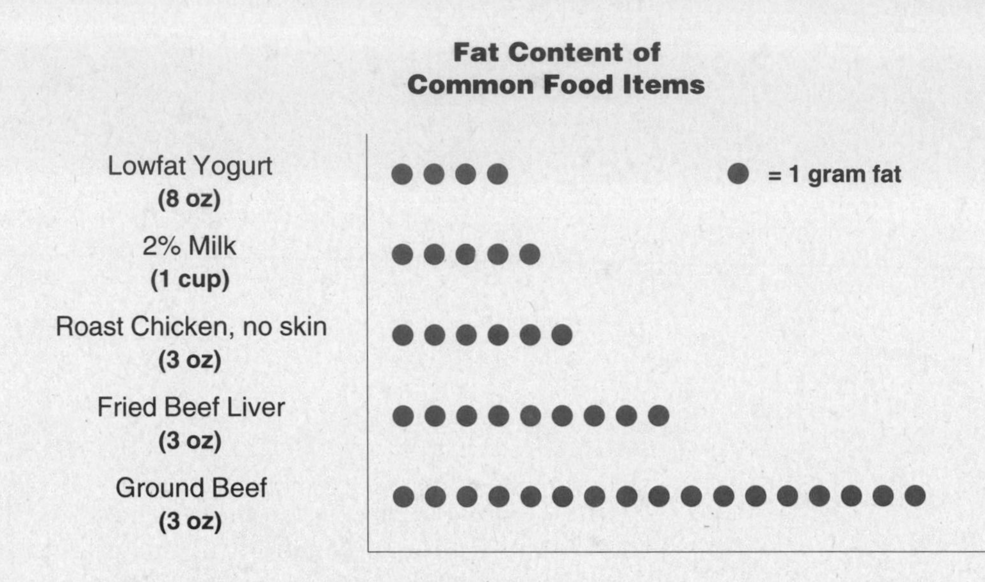

WN

1. How many fewer grams of fat are in 3 ounces of fried beef liver than in 3 ounces of ground beef?

(1) 8 (2) 9 (3) 17 (4) 24 (5) not enough information is given

2. Belinda drinks 3 cups of 2 percent milk every day. How many grams of fat does she consume in this milk?

(1) 5 (2) 10 (3) 15 (4) 20 (5) not enough information is given

3. Quentin ate 6 ounces of roast chicken with skin. How many grams of fat did the chicken contain?

(1) 3 (2) 4 (3) 5 (4) 6 (5) not enough information is given

Questions 4–7 refer to the chart below.

Indianapolis 500 Auto Race Winners

Year	Driver	Time	MPH (miles per hour)
1911	Ray Harroun	6 hrs 42 min 8 sec	74.59
1931	Louis Schneider	5 hrs 10 min 27.93 sec	96.629
1951	Lee Wallard	3 hrs 57 min 38.05 sec	126.244
1971	Al Unser	3 hrs 10 min 11.56 sec	157.735
1987	Al Unser	3 hrs 4 min 59.147 sec	162.175

4. How many miles per hour did the winning car travel in 1987?

(1) 59.147 (2) 74.59 (3) 162 (4) 162.175 (5) not enough information is given

5. To win the race, approximately how many times as long did the 1911 driver take than the 1971 driver?

(1) $\frac{1}{2}$ (2) 2 (3) 4 (4) 5 (5) not enough information is given

D

6. The slowest car in the 1971 Indianapolis 500 traveled at how many miles per hour?

(1) 74.59 (2) 96.629 (3) 126.244 (4) 157.735 (5) not enough information is given

7. About how many miles per hour faster did Al Unser travel in the 1987 race than in the 1971 race?

(1) 3 (2) 4 (3) 6 (4) 10 (5) not enough information is given

Answers begin on page 195.

MIXED REVIEW

Directions: Solve the following word problems.

Questions 1–4 refer to the graph below.

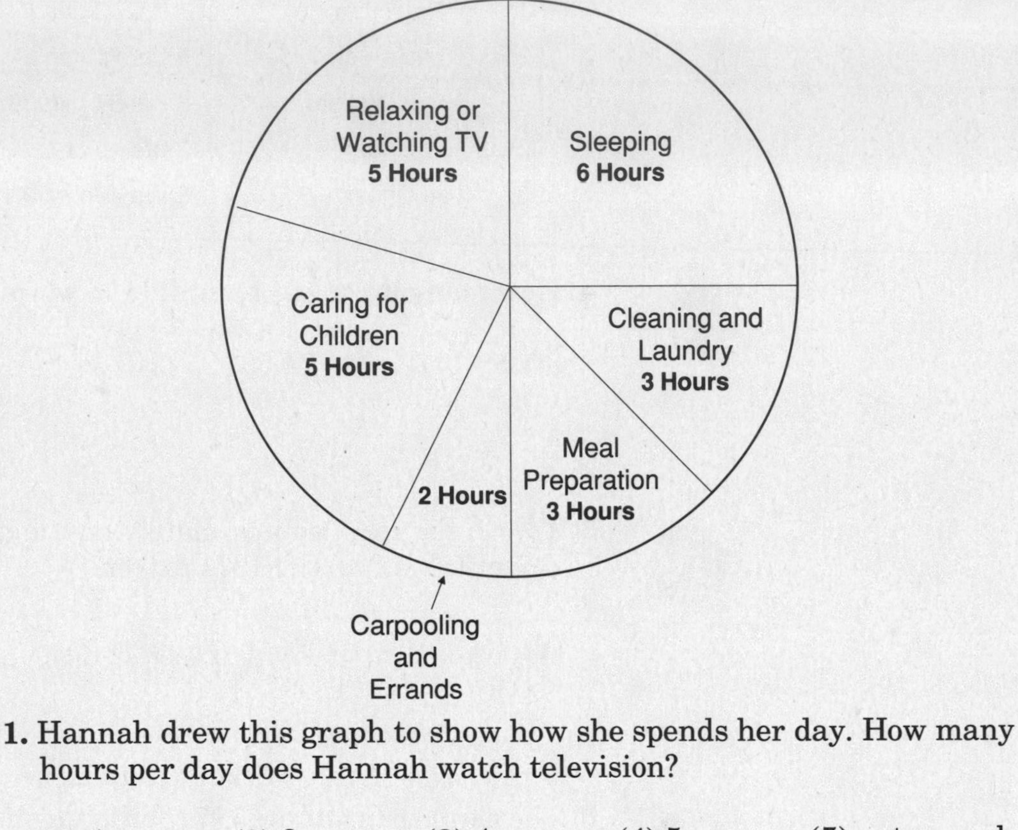

WN

1. Hannah drew this graph to show how she spends her day. How many hours per day does Hannah watch television?

(1) 1 (2) 2 (3) 4 (4) 5 (5) not enough information is given

F

2. Hannah feels she needs more sleep at night. If she were able to cut her cleaning and laundry time by $\frac{1}{3}$, how many total hours could she then spend sleeping?

(1) 1 (2) 6 (3) $6\frac{1}{3}$ (4) 7 (5) 8

3. What fraction of her day does Hannah spend preparing meals?

(1) $\frac{3}{1}$ (2) $\frac{1}{3}$ (3) $\frac{1}{4}$ (4) $\frac{1}{6}$ (5) $\frac{1}{8}$

P

4. What percent of her day does Hannah spend sleeping?

(1) 6 (2) 12 (3) 20 (4) 24 (5) 25

5. Floyd planned to make a ceramic decoration for each member of his family. Each decoration takes an hour to design, 2 hours to mold, and 1 hour to glaze. What more do you need to know to find out how long it will take Floyd to make all the decorations?

(1) the size of each decoration
(2) the amount Floyd will spend on materials
(3) the total hours that each decoration takes to make
(4) the number of people in Floyd's family
(5) the number of decorations Floyd can make in a day

Questions 6–8 refer to the following information.

While shopping, Jim saw this sign. He ordered 2 pounds of coleslaw, $1\frac{1}{2}$ pounds of roast beef, 3 pounds of bologna, and 1 pound of American cheese. He then went to the produce section, where he chose some carrots, greens, and tomatoes. On his way to the checkout Jim also chose a vegetable peeler for $3.20. He paid for all of his purchases with a $50 bill.

TODAY'S SPECIALS	
ROAST BEEF	$4.98 PER lb.
AMERICAN CHEESE	$2.28 PER lb.
BOLOGNA	$2.99 PER lb.
TURKEY BREAST	$3.99 PER lb.
POTATO SALAD	$2.10 PER lb.
COLE SLAW	$1.90 PER lb.

D

6. Jim's total purchases, including tax, came to $47.12. How much change should he receive?

(1) $1.10 (2) $2.88 (3) $31.90 (4) $97.12 (5) not enough information is given

7. In addition to receiving the special price on American cheese, Jim had a $.20-off coupon for this item. Which of the following expressions shows what he would pay for the cheese?

(1) $2.28 – $.20
(2) $\frac{1}{4}$ ($2.28) – $.20
(3) $\frac{1}{4}$ ($2.28 + $.20)
(4) 4($2.28 – $.20)
(5) $2.28 + $\frac{1}{4}$ (.20)

P

8. The vegetable peeler was subject to a 5 percent tax. What was the price of the peeler including tax?

(1) $.16 (2) $3.20 (3) $3.36 (4) $4.20 (5) not enough information is given

Answers begin on page 196.

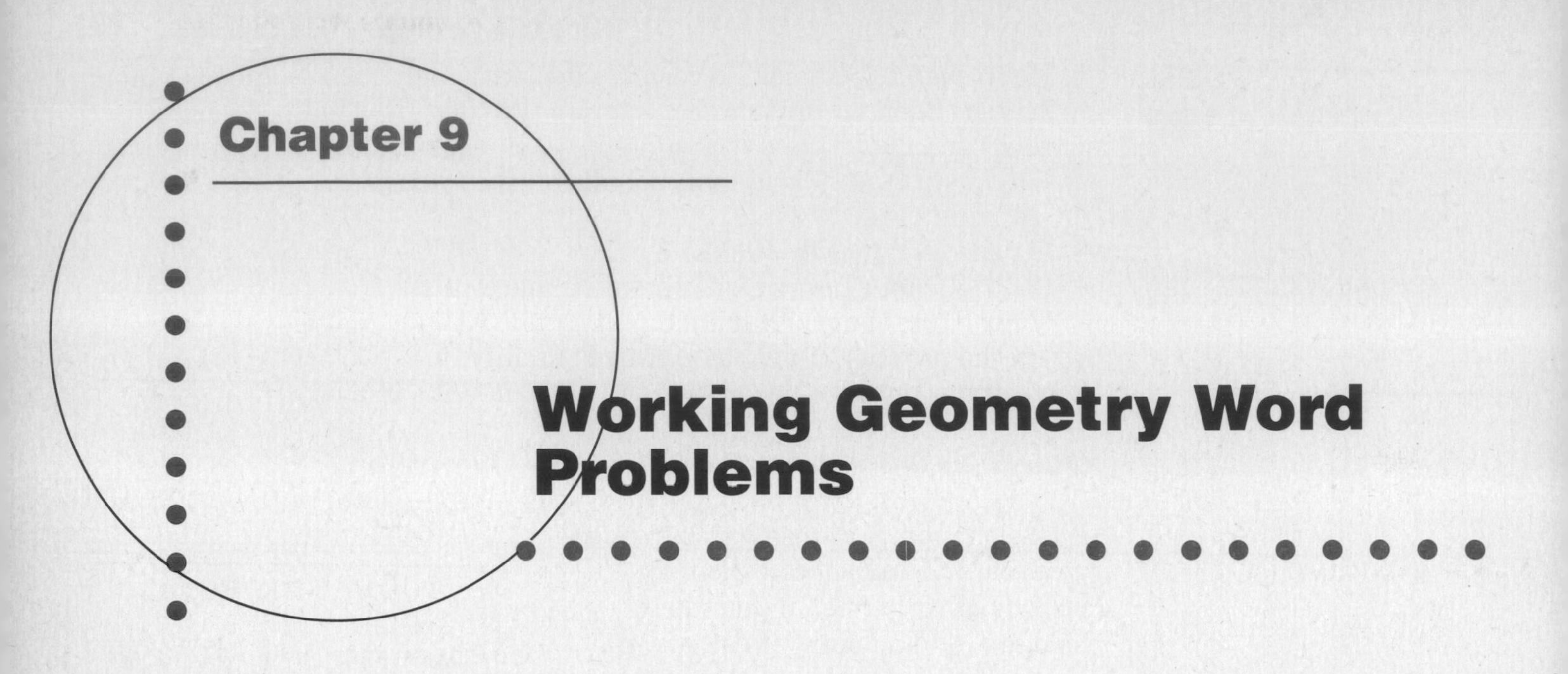

Chapter 9

Working Geometry Word Problems

FINDING INFORMATION ON DRAWINGS

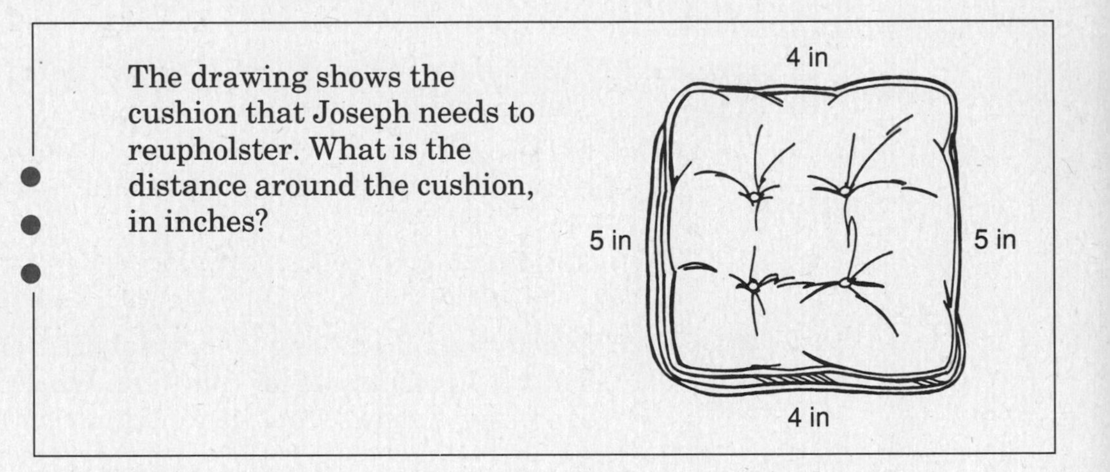

The drawing shows the cushion that Joseph needs to reupholster. What is the distance around the cushion, in inches?

How do you find the correct answer to this problem?

You probably know that by adding up the lengths of each of the sides of the figure you'll get the total distance around, called **the perimeter** of the figure.

$4 + 5 + 4 + 5 = 18$ inches

To get the correct answer you simply use the numbers written on the drawing—these numbers represent length.

Now look at another way that information can be represented on a drawing.

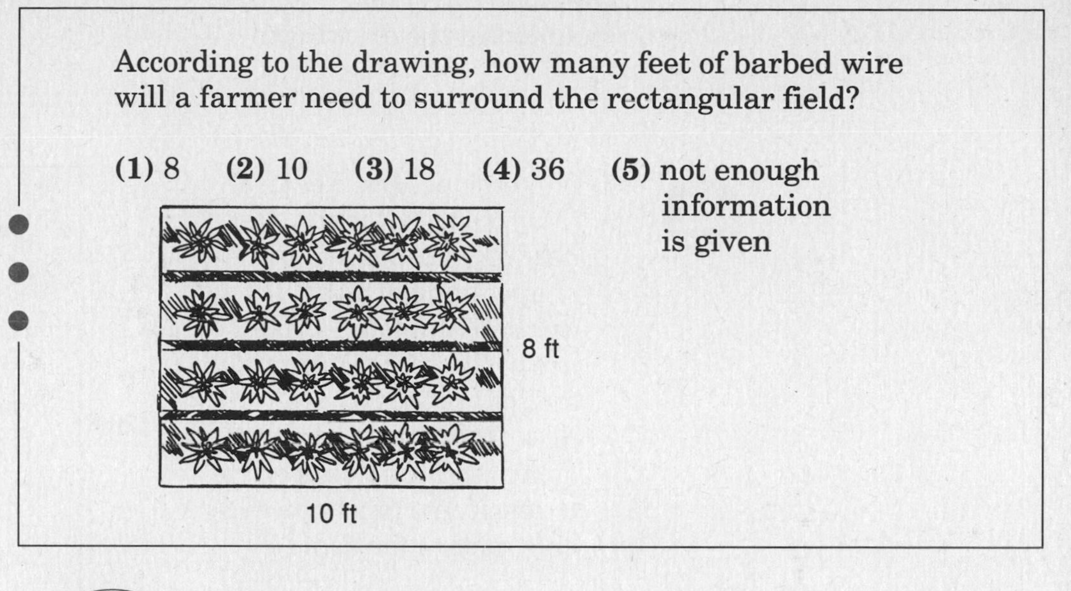

According to the drawing, how many feet of barbed wire will a farmer need to surround the rectangular field?

(1) 8 (2) 10 (3) 18 (4) 36 (5) not enough information is given

What is the length of each side of the rectangle?

How did you know?

At first it may seem that some information is left off the drawing. There are four sides to the figure, but only two lengths are labeled. Is the answer **(3) 18**? Or is it **(5) not enough information is given**?

To answer correctly you need to reread the problem. It states that the figure in the drawing is a *rectangle*. Your knowledge of geometry should remind you that in a rectangle the opposite sides are equal in length.

$$\underbrace{(2 \times 8)}_{\text{2 widths}} + \underbrace{(2 \times 10)}_{\text{2 lengths}} = 16 + 20 = 36$$

The correct answer is **(4) 36**.

If you are having trouble with a geometry problem, reread the problem AND look at the drawing again. You may have overlooked some information.

Exercise 1

Directions: Solve the following word problems using the information included in the drawings.

GE

1. A sketch of Mrs. Fein's floor plan is at right. Her bathroom is square in shape, and both the kitchen and family room are rectangles. How many feet of baseboard does Mrs. Fein need to buy to go around the floors of the three rooms (without subtracting for doors)?

8 ft
15 ft
8 ft
bath
kitchen
12 ft
20 ft
family room
8 ft

(1) 32 **(2)** 54 **(3)** 56 **(4)** 86 **(5)** 142

2. What is the distance, in centimeters, around the figure at right?

5 cm
5 cm
4 cm
4 cm

(1) 23 **(2)** 22 **(3)** 20 **(4)** 18 **(5)** not enough information is given

GE

3. How many bricks would a contractor need to buy to surround the yard pictured at right?

20 yd
25 yd
35 yd
30 yd

(1) 25 **(2)** 35 **(3)** 110 **(4)** 220 **(5)** not enough information is given

GE

4. If fabric costs $4.99 per square yard, approximately how much will Nina have to pay to cover just the front of the 3 frames shown below? (Remember: Area = length × width.)

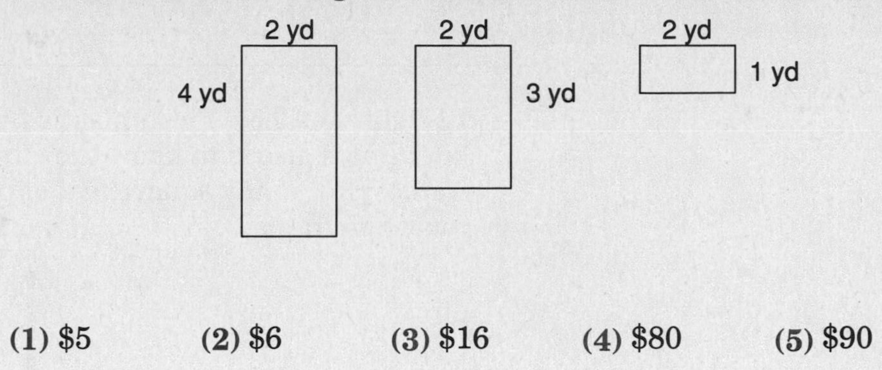

(1) $5 **(2)** $6 **(3)** $16 **(4)** $80 **(5)** $90

5. A teacher cut out 20 triangles from a piece of poster board. Each triangle had 3 equal sides as shown at right. What was the perimeter (in feet) of each triangle?

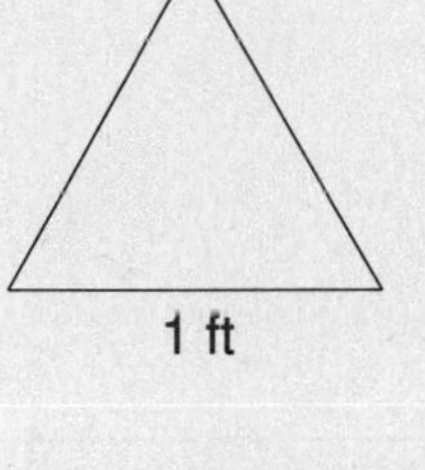

(1) 1 **(2)** 3 **(3)** 20 **(4)** 60 **(5)** not enough information is given

GE

6. How many yards of trim does Mr. Roscoe need to completely cover the perimeter of the rectangular window at right?

2 yd

(1) 2 **(2)** 4 **(3)** 6 **(4)** 8 **(5)** not enough information is given

Answers begin on page 196.

LET FORMULAS WORK FOR YOU

To figure out how much plastic he needs to make a new kite, Ralph needs to know the surface area of this wooden frame. How many square feet of plastic will he need to cover the frame?

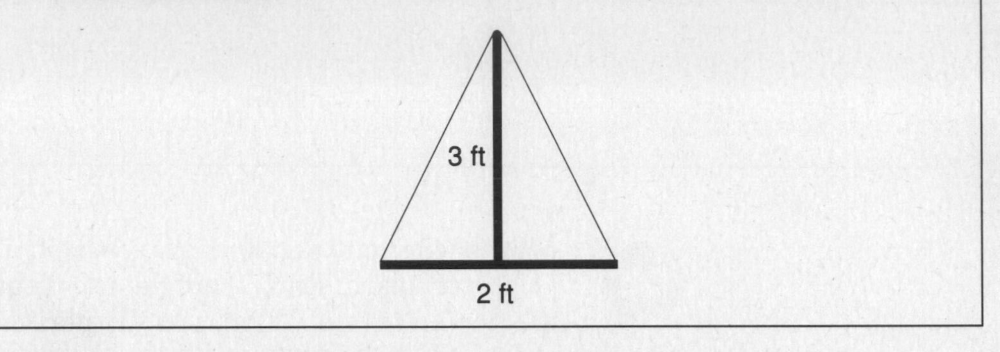

What is the shape of the kite Ralph is making?

What else do you need to know to answer this problem?

You probably can see that the kite pictured above is a **triangle** with a base of 2 feet and a height of 3 feet. To figure out how much plastic Ralph needs, you need to determine the **area** of the kite.

The AREA of a triangle = $\frac{1}{2}$ of the **base** times the height

The formula is written as: $A = \frac{1}{2}bh$

The equation above ($A = \frac{1}{2}bh$) is called a **formula**, and, like other formulas, it is extremely useful in solving geometry problems.

TIP

Multiplication is indicated in several ways in these formulas:

- a number and a letter—$4s$ means $4 \times$ side
- two letters—$l\,w$ means length × width
- squared or cubed—s^2 means side × side, and s^3 means side × side × side

To use a formula:

1. First write down the formula you need to use. • • For the problem, you'll use

   $A = \frac{1}{2}bh$

2. Then **substitute** any values that you know from the problem. • • • You know that the height is 3 and the base is 2.

   $A = \frac{1}{2}(2 \times 3)$
   ↑ base ↑ height

3. Now solve the equation. • • • • • • • • • • $A = \frac{1}{2}(2 \times 3)$

   $A = \frac{1}{2}(6)$ or **3 sq ft**

Remember that area is always expressed in square units.

NOTE: To refresh your memory about solving equations, turn back to page 86, where you first learned the rules.

FOR YOUR REFERENCE . . .

- A complete list of formulas is on page 165. You may use this list for reference as you work through this chapter.

Exercise 2

Directions: Solve the following problems. First write the formula you will use. (If you need to, use the formulas on page 165.) Then solve the problem.

Example: Sally needs to surround a square piece of wood with a plastic border. How many inches of plastic will she need for the perimeter?

Formula: $P = 4s$

Answer: $P = 4s$
$P = 4 \times 8''$ ←1 side
$P = 32''$

1. An artist begins a sculpture by forming circumferences for 3 wire circles of the size shown below. Which expression shows the number of inches of wire he will need?

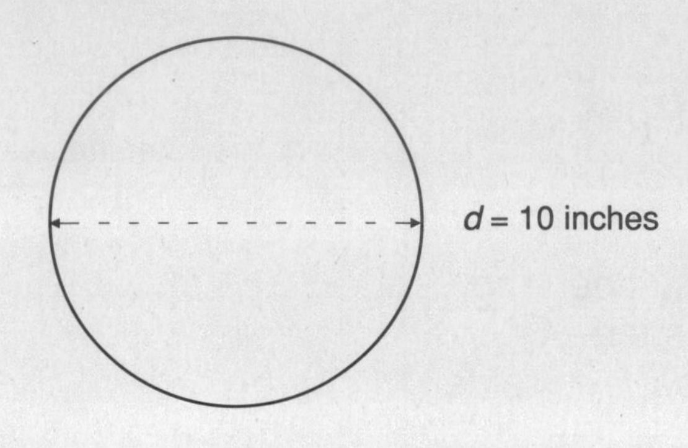

(1) 3.14×10
(2) $3\,(3.14 \times 10)$
(3) $3.14 \times 10 \times 10$
(4) $3\,(10 \times 10 \times 3.14)$
(5) not enough information is given

Formula:

Answer:

2. Using the measurements shown here, how many *yards* of fencing will a dog owner need to surround the perimeter of a dog pen?

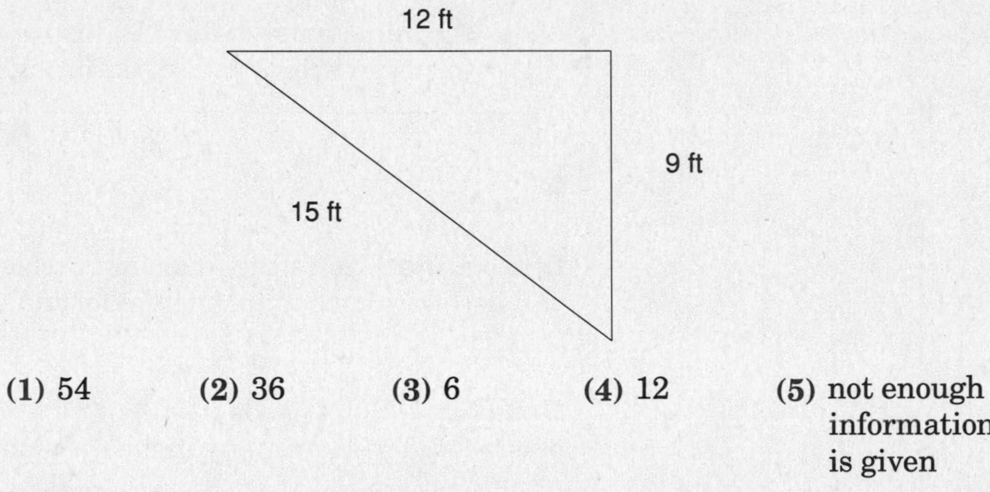

(1) 54 (2) 36 (3) 6 (4) 12 (5) not enough information is given

Formula:

Answer:

3. What is the volume of cement, in cubic centimeters, that can be held in the cube below?

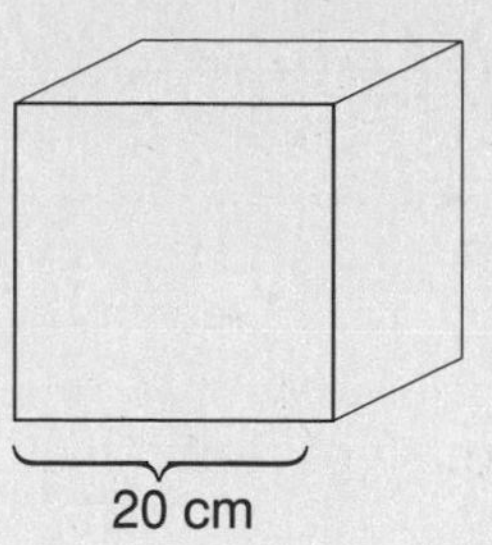

(1) 80 **(2)** 400 **(3)** 8,000 **(4)** 160,000 **(5)** not enough information is given

Formula:

Answer:

4. Glowing Floor Company charges \$.50 per square foot to clean and wax wood floor surfaces. Which of the following expressions shows the amount the company would charge to clean and wax the entire wood floor shown below?

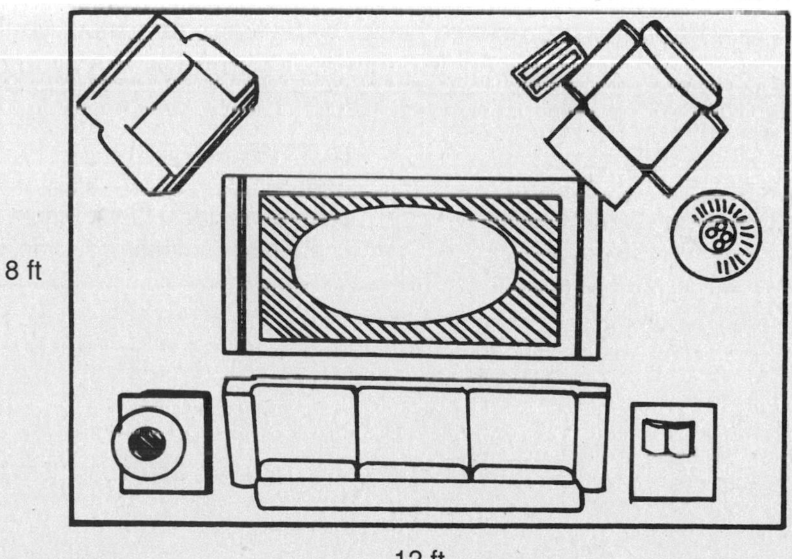

(1) .50 (8 + 12)
(2) $\frac{1}{2}$ (.50 × 8 × 12)
(3) .50 (8 × 12)
(4) .50 (8 + 8 + 12 + 12)
(5) not enough information is given

Formula:

Answer:

Answers begin on page 196.

VISUALIZING GEOMETRY PROBLEMS

The canister shown at right is filled to the top with oat flour. How many cubic inches of flour does it contain?

h = 6 inches, r = 2.5 inches

Look at the list of formulas on page 165. Which formula should you use to solve this problem?

How did you decide?

Fortunately, you don't have to memorize hundreds of formulas and be able to recall them on a moment's notice. You can look in math or reference books to find the formulas you need. In fact, many kinds of tests include a list of formulas that you can use for handy reference.

BUT REMEMBER!
Even when you have a list of formulas, you still need to decide *which* formula is the correct one to use. This is sometimes a tricky issue.

In the problem above there are some clues that tell you which formula to use:

1. The shape shown in the drawing is a *cylinder*.

2. The problem asks for the number of *cubic* inches.

Only the formula for the *volume of a cylinder* will work in this case.

$V = \pi r^2 h$

$V = 3.14 \times (2.5)^2 \times 6$

$V = 3.14 \times 6.25 \times 6$

V = **117.75 cubic inches**

CLUES FOR DECIDING FORMULAS . . .

1. What kinds of units are being used?

- *Cubic* units are used in *volume* problems.
- *Square* units are used in *area* problems.
- *Regular* units are used in *perimeter* problems.

2. What is the shape of the figure in question?

- A *flat* figure means *area* or *perimeter* (or *circumference* of a circle).
- A *three-dimensional* or *solid* figure means *volume*.

3. Can you picture in your mind what is happening in the problem?

- A problem concerning a **surface**, such as carpeting, tile, fertilizer, fabric, or soil, most often is asking for *area*.
- A problem concerning **distance around** an object, such as fencing, rope, or weather stripping that surrounds a shape, most often is asking for *perimeter*.
- A problem that concerns how much a figure **holds** most often is asking for *volume*.

REMEMBER!
There are other formulas that are not geometry formulas. They are included in the list on page 165, and you should get used to using these formulas as well.

Exercise 3

Directions: Use the list of formulas on page 165 to solve the following problems. First circle the kind of space you will be finding. Write what clue you used. Then write the formula and solve the problem.

Example: Max and Bertha Mahoney were responsible for roping off a children's play area at the school fair. The diameter of the circle they planned was 21 feet. How many feet of rope will they need? (Use $\frac{22}{7}$ for π.)

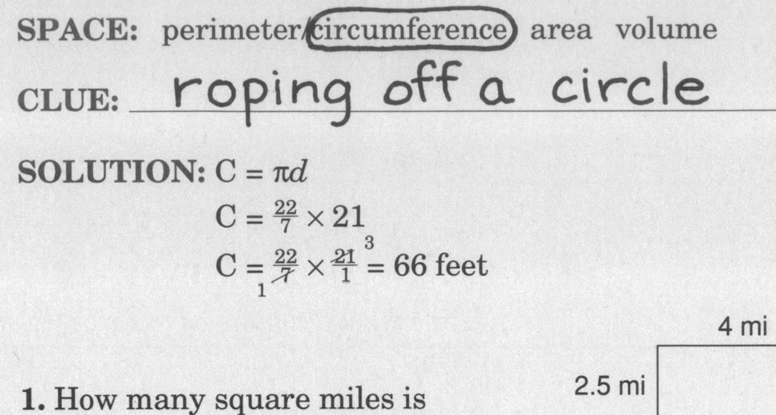

SPACE: perimeter/circumference area volume

CLUE:

SOLUTION: $C = \pi d$

$C = \frac{22}{7} \times 21$

$C = \frac{22}{\not{7}_1} \times \frac{\not{21}^3}{1} = 66$ feet

GE **1.** How many square miles is the plot of land pictured here?

4 mi

2.5 mi

SPACE: perimeter/circumference area volume

CLUE:

SOLUTION:

GE **2.** Every cubic foot of material inside the rectangular container weighs 4.5 pounds. If the box is completely filled, how many pounds of material does it contain?

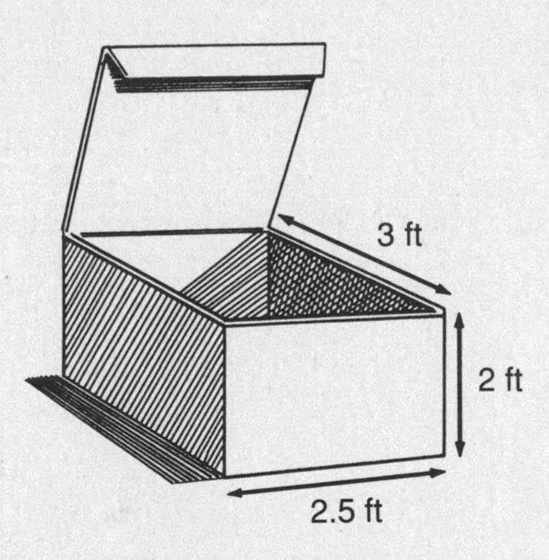

SPACE: perimeter/circumference area volume

CLUE:

SOLUTION:

3. A seamstress bought 12 feet of gold braid to surround this scarf. How long is the third side of the scarf?

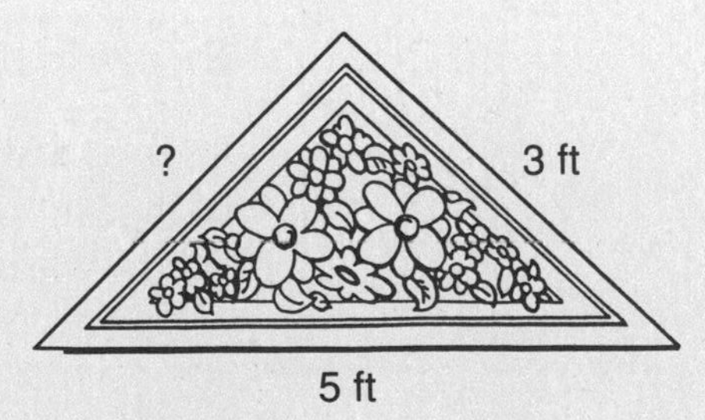

SPACE: perimeter/circumference area volume

CLUE:

SOLUTION:

Answers begin on page 197.

Formulas

Description	Formula
AREA (A) of a:	
square	$A = s^2$; where s = side
rectangle	$A = lw$; where l = length, w = width
parallelogram	$A = bh$; where b = base, h = height
triangle	$A = \frac{1}{2}bh$; where b = base, h = height
circle	$A = \pi r^2$; where $\pi = 3.14$, r = radius
PERIMETER (P) of a:	
square	$P = 4s$; where s = side
rectangle	$P = 2l + 2w$; where l = length, w = width
triangle	$P = a + b + c$; where a, b, and c are the sides
circumference (C) of a circle	$P = \pi d$; where $\pi = 3.14$, d = diameter
VOLUME (V) of a:	
cube	$V = s^3$; where s = side
rectangular container	$V = lwh$; where l = length, w = width, h = height
cylinder	$V = \pi r^2 h$; where $\pi = 3.14$, r = radius, h = height
Pythagorean relationship	$c^2 = a^2 + b^2$; where c = hypotenuse, a and b are legs of a right triangle
distance (d) between two points in a plane	$d = \sqrt{(x_2 - x_1)^2 + (y_2 - y_1)^2}$; where (x_1, y_1) and (x_2, y_2) are two points in a plane
slope of a line (m)	$m = \frac{y_2 - y_1}{x_2 - x_1}$; where (x_1, y_1) and (x_2, y_2) are two points in a plane
mean	$mean = \frac{x_1 + x_2 + \ldots x_n}{n}$; where the x's are the values for which a mean is desired, and n = number of values in the series
median	*median* = the point in an ordered set of numbers at which half of the numbers are above and half of the numbers are below this value
simple interest (i)	$i = prt$; where p = principal, r = rate, t = time
distance (d) as function of rate and time	$d = rt$; where r = rate, t = time
total cost (c)	$c = nr$; where n = number of units, r = cost per unit

Source: GED Testing Service of the American Council on Education, © 1987

PICTURING A GEOMETRY PROBLEM

Peter and Lee had driven 18 miles south when they discovered they were lost. They then drove 24 miles east to get to Medfield. What is the shortest distance (in miles) between Medfield and their starting point?

It is often helpful to draw a rough picture of the things taking place in a problem. This may help you to "see" a solution more easily.

How do you solve this problem? Is it a geometry problem or a simple distance problem?

If you read through this problem quickly, you might be tempted to find 42 as the answer (18 + 24 = 42). However, notice that the question does *not* ask for the total distance traveled. Instead it asks for the *shortest* distance between two points. This problem is, in fact, a geometry problem.

Here are some helpful steps in solving a problem like the one above:

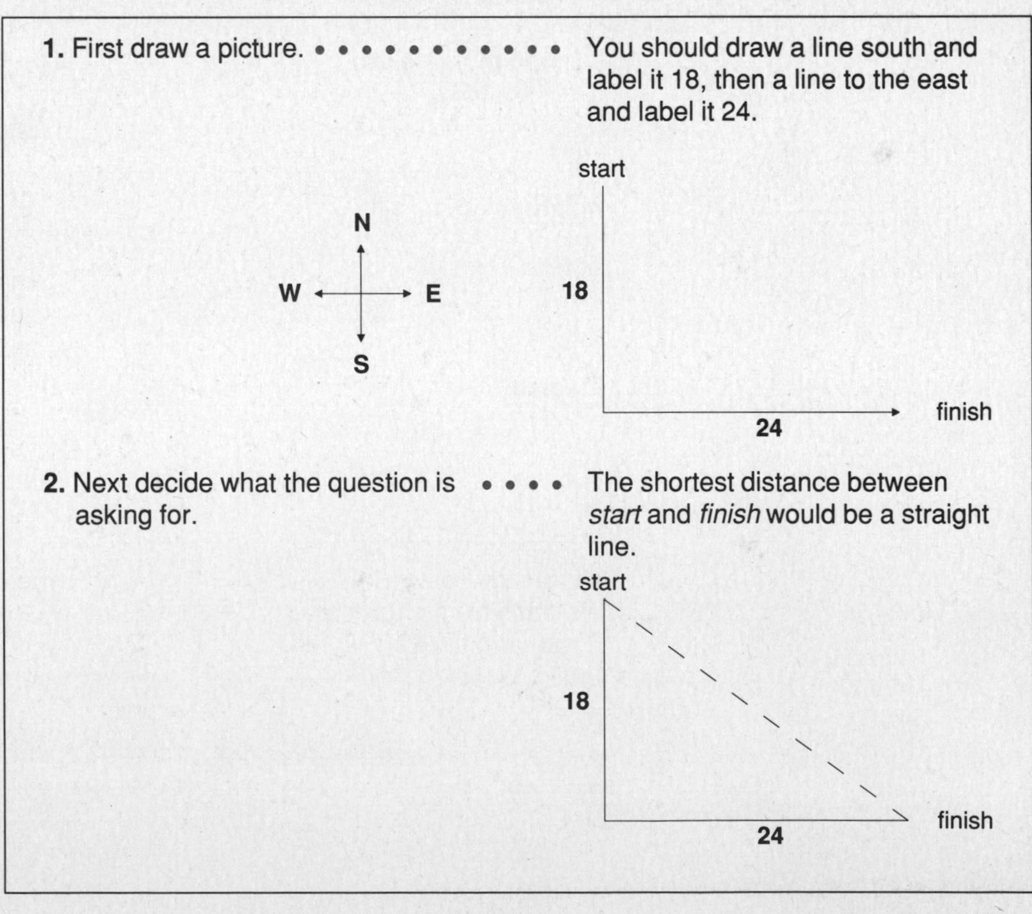

1. First draw a picture. • • • • • • • • • • • You should draw a line south and label it 18, then a line to the east and label it 24.

2. Next decide what the question is asking for. • • • • The shortest distance between *start* and *finish* would be a straight line.

3. Once you can "see" the geometry problem, choose a formula that will help you.	This is a picture of a right triangle. I can use the formula $a^2 + b^2 = c^2$ to find the length of the side (the **hypotenuse**).
4. Substitute the values that you know, then solve.	$c^2 = a^2 + b^2$ $c^2 = 18^2 + 24^2$ $c^2 = 324 + 576$ $c = \sqrt{900}$ $c = 30$

Now let's look at another geometry problem that might be hard to recognize.

How many 1-yard-square rubber pads will be needed to cover the playground area shown below?

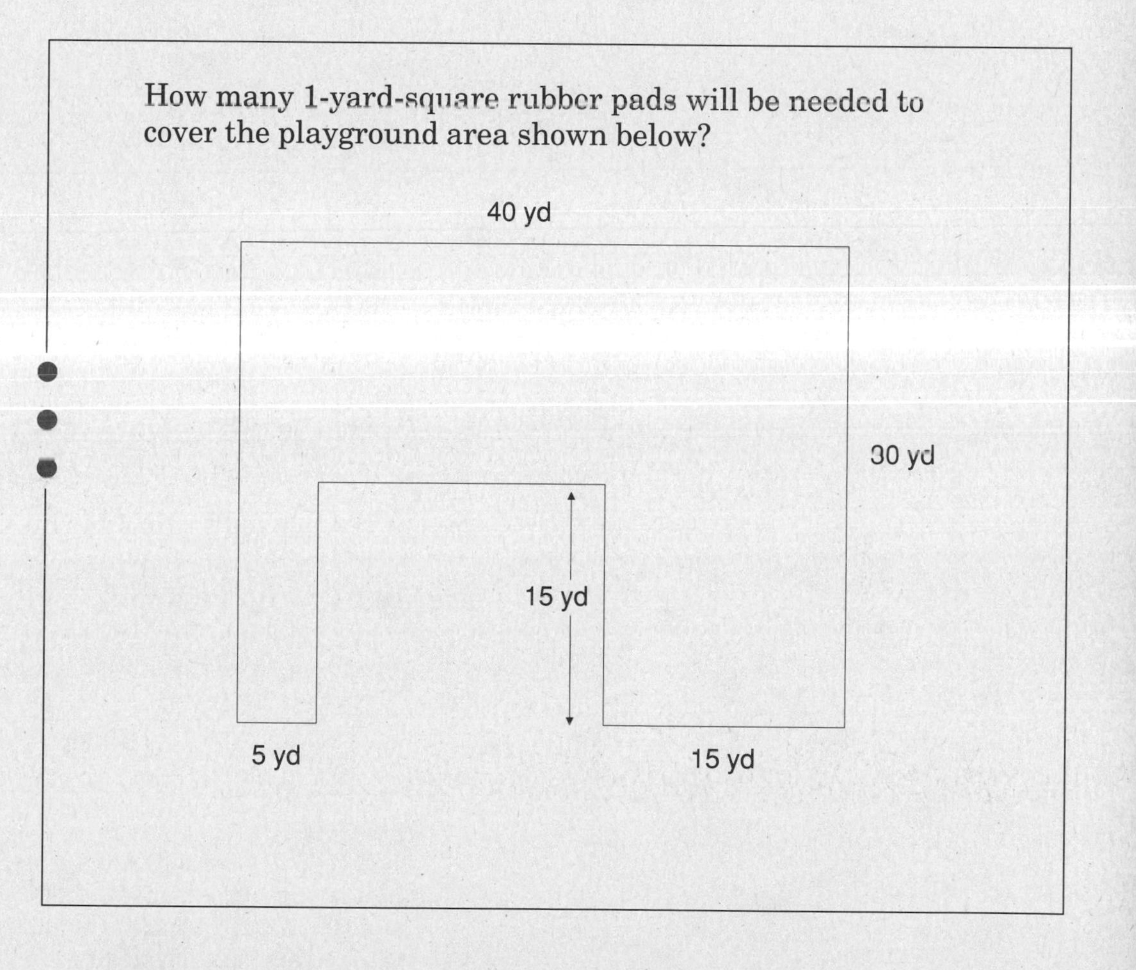

Is there a formula to determine the area of a shape like this?

How can you find the area?

Don't give up if you find an unfamiliar shape like the one on page 167! Instead, see if the shape can be broken up into smaller figures that you *do* recognize.

Here is the same shape, divided into smaller areas:

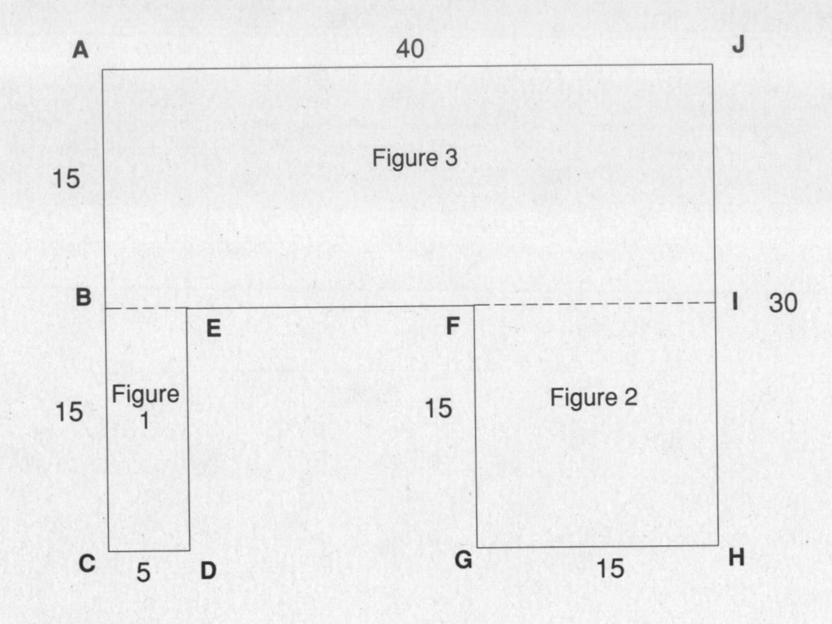

As you can see:

- Figure 1 (BCDE) is a rectangle 5 yd by 15 yd.
- Figure 2 (FGHI) is a square 15 yd by 15 yd.
- Figure 3 (ABIJ) is a rectangle 15 yd by 40 yd.

To find the area of the whole figure, add the areas of the smaller figures:

Figure 1 Figure 2 Figure 3

$(5 \times 15) + (15 \times 15) + (15 \times 40) =$

$75 + 225 + 600 =$ **900 sq yd**

> **REMEMBER!**
> When you have broken up the shape into smaller, more recognizable shapes, you may have to do some subtraction to find the lengths of some sides. For example, if you know that side JH is 30 yd and FG and HI are both 15 yd, you can subtract (30 – 15) to find IJ.

There can be several different ways to break up a shape like the one above. Simply look for squares, rectangles, or triangles in the figure and try different ways of breaking the larger figure into smaller shapes.

Exercise 4

Directions: Solve the following problems. Remember to draw a picture if you need help deciding what to do.

Questions 1–3 refer to the following information.

The Howes want to redo the floors in their home. The floor plan is pictured below. The carpeting they like is $8.95 per square yard installed, and the linoleum they like is $.75 per square foot if they install it themselves.

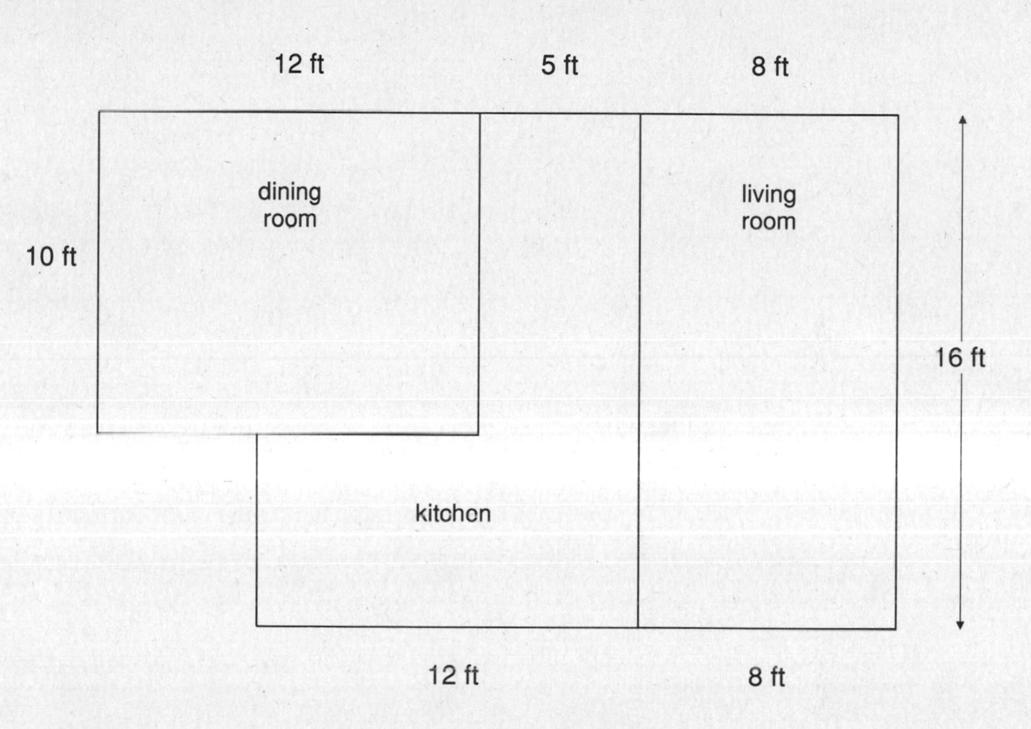

GE **1.** If the Howes decide to carpet their living room and dining room, how many square yards of carpeting do they need? (Round to the nearest square yard.)

GE **2.** If they carpet both the living room and dining room, how much will the carpet cost before tax?

GE **3.** How many square feet of linoleum will the Howes need to cover the kitchen?

GE **4.** A trucker drove 160 miles north from Freeport, then drove 120 miles west to Dalton. What is the shortest distance in miles from Freeport to Dalton?

Answers begin on page 197.

MIXED REVIEW

Directions: Solve the following word problems.

1. Which of the following expressions shows the area of a circle with a radius of 5?

(1) 5×5
(2) $(3.14) \times 5 \times 2$
(3) 3.24×5
(4) 3.14×5^2
(5) not enough information is given

D

2. A mother was comparing two bottles of pain relievers at the grocery store. One bottle was $2.40 for 50 tablets; the other was $3.00 for 75 tablets. What is the difference in price *per tablet*?

(1) $.002 (2) $.008 (3) $.02 (4) $.20 (5) $.60

P

3. The Gold Company charges 7.5 percent monthly interest on any unpaid balance on its credit cards. In September, Mary bought a dress for $49 and a toaster oven for $79 with her credit card. When she received her bill, she paid $128. What interest does she owe?

(1) $128.00 (2) $118.40 (3) $9.60 (4) $4.40 (5) 0

Questions 4–6 refer to the following graph.

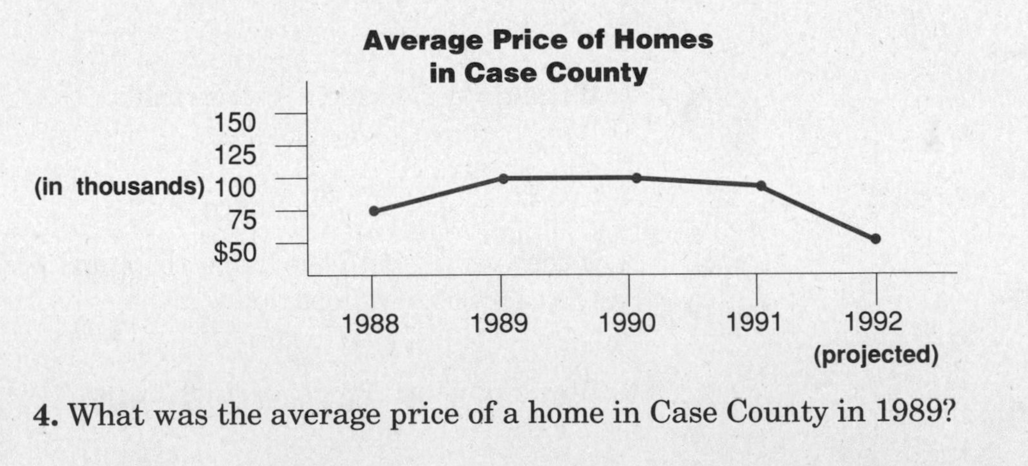

4. What was the average price of a home in Case County in 1989?

(1) $75 (2) $100 (3) $75,000 (4) $100,000 (5) $175,000

5. By approximately how many dollars did the price of a Case County home fall between 1990 and 1991?

(1) $100,000 (2) $90,000 (3) $25,000 (4) $10,000 (5) $10

6. According to the graph, what is the projected 1992 price of a Case County home?

(1) \$75,000 (2) \$50,000 (3) \$25,000 (4) \$50 (5) not enough information is given

GE

7. A triangular piece of metal has a total surface area of 57 square feet. The base of the piece measures 6 feet across. Which expression shows how many feet tall the piece is?

(1) $h = 57 \times (\frac{1}{2} \times 6)$
(2) $h = 57 \div (\frac{1}{2} \times 6)$
(3) $h = 57 \div 6$
(4) $h = \frac{1}{2}(57 \times 3)$
(5) not enough information is given

P

8. A bank customer earned \$44.64 interest in one year for a deposit he made in a savings account. If the bank pays 8% yearly interest, how much money had the customer deposited?

(1) \$3.57 (2) \$357.12 (3) \$558.00 (4) \$600.00 (5) not enough information is given

GE

9. An aluminum can holds 19.85 cubic centimeters of liquid. What more do you need to know to find the height of the can?

(1) the total weight of the filled can
(2) the weight of the empty can
(3) the volume of the can
(4) the length of a side of the can
(5) the radius of the can

10. Last month Marcel, an artist, sold a sculpture for \$400.00, an oil painting for \$797.50, and a watercolor for \$297.50. Approximately what fraction of these earnings does the watercolor represent?

(1) $\frac{1}{2}$ (2) $\frac{1}{3}$ (3) $\frac{1}{4}$ (4) $\frac{1}{5}$ (5) not enough information is given

Answers begin on page 198.

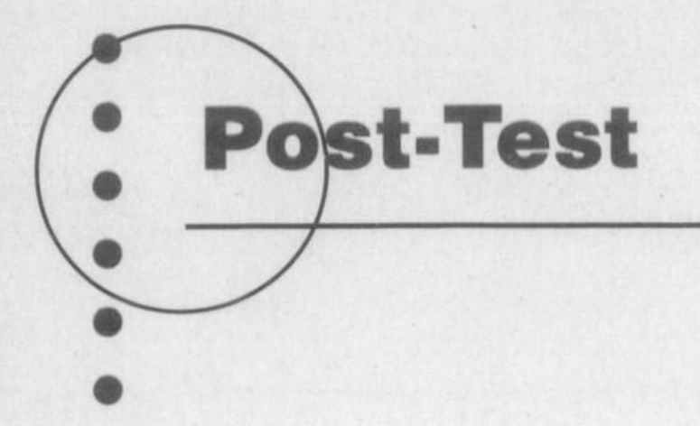

Post-Test

Directions: Solve the following word problems. Choose the best answer from the choices given. **If you need to, you may refer to the formulas on page 165.**

1. Antonio mixed 4 pints of melon, 3 pints of strawberries, and 6 pints of peaches together in a salad. He then divided the salad into 12 servings for his guests. Which of the following expressions shows the amount of salad each guest received, in pints?

(1) $\frac{4 \times 3 \times 6}{12}$
(2) $(12 \div 3) + (12 \div 4) + (12 \div 6)$
(3) $12 - (4 + 3 + 6)$
(4) $\frac{4 + 3 + 6}{6}$
(5) $\frac{4 + 3 + 6}{12}$

Items 2–4 refer to the following chart.

Masciave Bakery
Weekly Production Report

Day	Usable Pies Produced	Number Rejected
Sunday	75	15
Monday	30	4
Tuesday	30	2
Wednesday	40	4
Thursday	60	10
Friday	100	10
Saturday	100	4

2. According to the chart, what is the average number of pies rejected per day at the Masciave Bakery?
(1) 7
(2) 49
(3) 62
(4) 386
(5) 435

3. How many more usable pies are produced on Thursday than on Wednesday?

(1) 10
(2) 20
(3) 60
(4) 100
(5) not enough information is given

4. If each usable pie sells for $9.95, and each rejected pie sells at a discount for $4.00, which of the following expressions shows the amount of money the bakery takes in for pies on Monday?

(1) 34 (9.95 + 4.00)
(2) 30 (4.00) + 4 (9.95)
(3) 30 (9.95) + 4 (4.00)
(4) (30 − 4) × (9.95 + 4.00)
(5) (30 + 4) × (9.95 − 4.00)

5. How many felt squares are needed to complete the quilt shown?

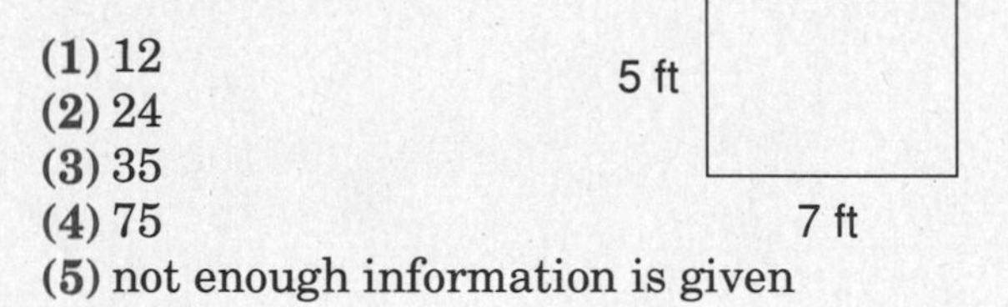

(1) 12
(2) 24
(3) 35
(4) 75
(5) not enough information is given

6. If each routine examination takes 25 minutes, approximately how many of these exams can a dentist fit in if she works for 6 hours straight?

(1) 1
(2) 6
(3) 12
(4) 30
(5) not enough information is given

7. Mrs. Vanderhagen canned 12 jars of plums, each containing 2 pints. She then canned 5 1-pint jars of plums. Which of the following expressions will tell you how many pints she canned in all?

(1) $12 + 5$
(2) $2\,(12 + 5)$
(3) $(2 \times 12) + 5$
(4) $(2 \times 5) + 12$
(5) $(12 + 2) + (5 + 1)$

8. A medium-size U-Pick-Up vehicle can haul 500 pounds of gravel. Tim is working a job in which he needs to haul 2400 pounds from the quarry to the worksite. If he rents the medium-size truck, what is the fewest number of trips he needs to make?

(1) 3
(2) 4
(3) 5
(4) 6
(5) not enough information is given

9. Eight people can stuff 960 envelopes in 4 hours. At this rate, how many envelopes can the group stuff in 5 hours?

(1) 150
(2) 600
(3) 1,000
(4) 1,200
(5) 1,500

10. Eddie bought a half-dozen donuts at $.49 each and two large coffees at $.75 each. How much money did he spend in all?
(1) $2.94
(2) $4.44
(3) $5.88
(4) $7.38
(5) not enough information is given

Items 11–14 refer to the following graph.

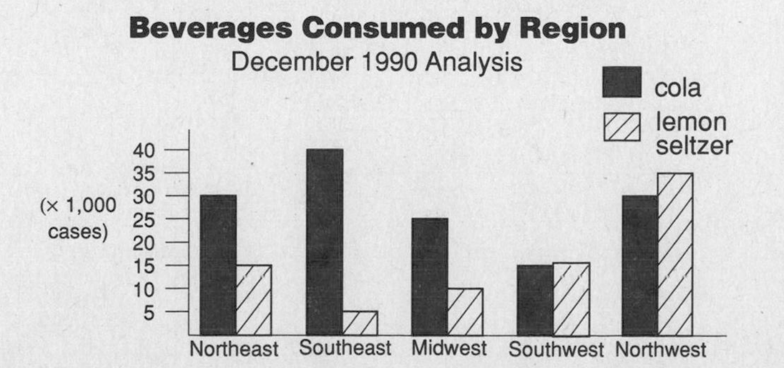

11. In what region was the consumption of cola and lemon seltzer about the same?

(1) Northeast
(2) Southeast
(3) Midwest
(4) Southwest
(5) Northwest

12. About how many more cases of cola than lemon seltzer were consumed in the Southeast in December 1990?

(1) 35
(2) 40
(3) 45
(4) 35,000
(5) 40,000

13. Half of all cola consumed in the Northwest in December 1990 was sold in metropolitan areas. How many cases of cola were sold in metropolitan areas?

(1) 15
(2) $17\frac{1}{2}$
(3) 15,000
(4) 17,500
(5) 32,500

14. During the first month of 1991, sales for diet beverages rose above sales for all colas and seltzers in the Northeast. Approximately how many cases of diet beverages were sold?

(1) 45
(2) 46
(3) 45,000
(4) 46,000
(5) not enough information is given

---

15. How many square yards of plastic are needed to cover the box shown here?

(1) 5
(2) 9
(3) 45
(4) 135
(5) not enough information is given

9 ft
5 ft

16. Mrs. Soo took her Girl Scout troop on a hike through the county forest. They followed a one-mile (5,280-foot) circular path. About how many feet did they then have to walk to get to the very center of the circle? (HINT: Use 3 for the value of π when you estimate.)

(1) 31,680
(2) 1,760
(3) 880
(4) 440
(5) not enough information is given

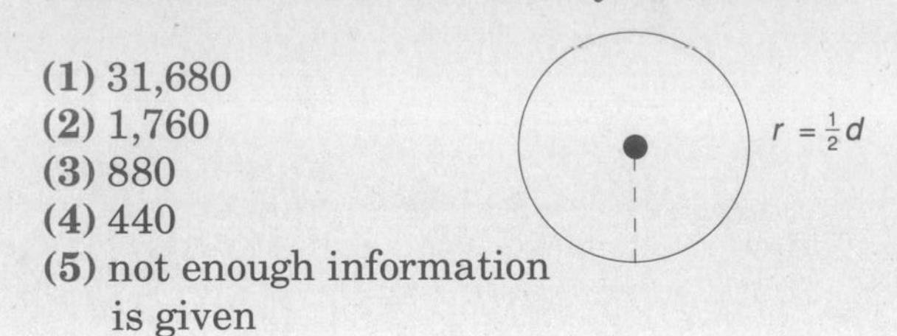

17. Find the correct order in weight of the blocks shown, listing them from heaviest to lightest.

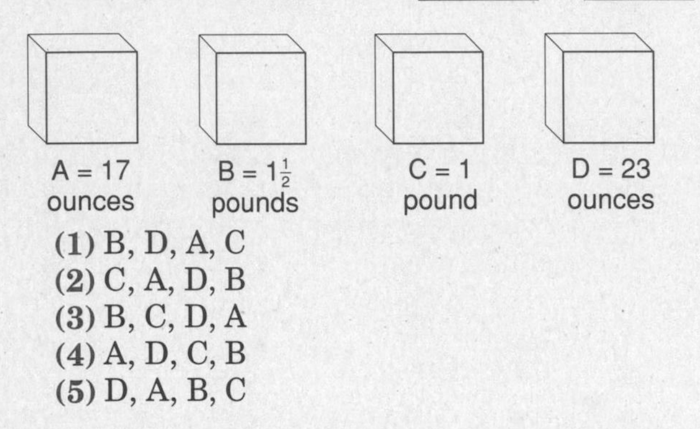

(1) B, D, A, C
(2) C, A, D, B
(3) B, C, D, A
(4) A, D, C, B
(5) D, A, B, C

18. For his English class, John wrote a 42-page paper, a 10-page paper, and a 16-page paper. Which of the following expressions shows the average page length of John's two longest papers?

(1) $\frac{42 + 10 + 16}{3}$
(2) $\frac{42 + 10 + 16}{2}$
(3) $\frac{42 + 16}{2}$
(4) $\frac{42 + 16}{3}$
(5) $42 - 16$

19. Grace Anne worked 30 hours at her regular pay of $5.50 per hour. She also worked 8 hours of overtime. What more do you need to know to find out Grace Anne's rate of overtime pay?

(1) her gross pay for the 38 hours worked
(2) her total monthly hours
(3) her yearly salary before taxes
(4) the amount of her social security, state, and federal taxes
(5) her total wages for the 30 regular hours

20. Virginia mixed together 15 ounces of lemonade, 80 ounces of tonic, and 20 ounces of juice. She then poured out an 8-ounce drink. How many ounces of the mixture were left in the pitcher?

(1) 115
(2) 107
(3) 105
(4) 95
(5) 92

21. The distance around a square field is 540 yards. How many yards is one side of the field?

(1) 35
(2) 40
(3) 135
(4) 140
(5) not enough information is given

Items 22–25 refer to the following information.

Paul and Kathleen must decide where to hold their parents' anniversary party. The Knights of Columbus charges $100 per hour for rental of its hall, and food and drink can be catered or brought in. The local social club charges $75 per hour for hall rental, but food and drink must be purchased through the club. A fee of $17 per guest is charged for a buffet (including beverages).

If Paul and Kathleen decide to rent the Knights of Columbus hall, they can order party platters from their local deli. Paul's brother owns a grocery store, and they can buy beverages at cost there. The charts below show the prices they would have to pay.

Party Platters (each serves 25 people)		
Assorted cold cuts	—	$19.95
Breads and rolls	—	8.00
Vegetables and dip	—	12.50
Crackers and cheese	—	12.50

Beverages		
1 quart bottled water	—	$1.19
Case (24 cans) soda	—	8.80
1 quart juice	—	1.45

22. How much would Paul and Kathleen pay for the party, including hall rental and buffet, if they had it at the local social club for 4 hours?

(1) $75
(2) $92
(3) $300
(4) $317
(5) not enough information is given

23. If they rent the Knights of Columbus hall, Paul and Kathleen have a budget of $400 for food and beverages. They will need 4 cold cut platters, 4 bread platters, 4 vegetable platters, and 4 crackers and cheese platters. How much money will they have left to spend on beverages?

(1) $200
(2) $211.80
(3) $188.20
(4) $88.20
(5) not enough information is given

24. What would the total cost of the party be if 75 guests were invited for 3 hours at the local social club?

(1) $225
(2) 1,275
(3) 1,500
(4) 1,700
(5) not enough information is given

25. Paul estimates that their guests would drink an average of $\frac{1}{2}$ quart of juice each. If 96 guests come to the party, how much will Paul need to spend on juice?

(1) $1.45
(2) 57.12
(3) 69.60
(4) 139.20
(5) 422.40

---

26. Taya would like to enlarge the photograph below. If the new depth she wants is 14 inches, how many inches wide would it be?

(1) $2\frac{1}{2}$
(2) 5
(3) 10
(4) 12
(5) 14

7 in

5 in

27. Kerry's shopping spree resulted in a sundress for $45.00, a cookbook for $9.95, a romance novel for $5.95, and a children's book for $8.99. She received $.11 in change. How much did Kerry spend on books alone?

(1) $70.00
(2) 69.89
(3) 25.00
(4) 24.89
(5) 24.78

28. A seamstress charges $8 per hour for hemming men's trousers. On an average day she hems 25 pairs. What more do you need to know to find the amount of money the seamstress earns on an average day?
(1) her yearly earnings
(2) the length of her lunch hour
(3) the cost of her materials
(4) the time of day she starts work
(5) the amount of time it takes to hem one pair of pants

Items 29 and 30 refer to the following circle graph.

29. How many minutes are spent in cleanup each day at the Newton Day-Care Center?
(1) $\frac{1}{2}$
(2) 5
(3) 6
(4) 30
(5) not enough information is given

30. The center charges $95 per week per child for 5 full days. Based on the hours shown on the graph, *approximately* how much do parents pay per hour for this child care?

(1) $1.00
(2) $2.00
(3) $4.00
(4) $20.00
(5) not enough information is given

EVALUATIVE POST-TEST CHART

Directions: Circle any item number that was answered *incorrectly*. Go back to the chapter numbers indicated and review the skill areas in which you had difficulty.

ITEM NUMBER	SKILL AREA	CHAPTER NUMBERS
10,20	Multi-Step Problems	1
1,4,7,18	Set-Up Questions	3,6
15	Understanding the Question	3
19,28	What More Do You Need to Know?	3
14	Not Enough Information	4
22,23,24,25	Working with Item Sets	4
5,27	Extra Information	4
9,26	Writing Proportions for Word Problems	5
6,30	When an Estimate Is the Answer	6
8	What to Do with Remainders	6
17	Comparing and Ordering	6
2,3,11,12,13,29	Using Charts, Graphs, and Drawings	8
16,21	Picturing a Geometry Problem	9

POST-TEST ANSWERS

1. (5) $\dfrac{4+3+6}{12}$

2. (1) 7
$15 + 4 + 2 + 4 + 10 + 10 + 4 = 49$
$49 \div 7$ days $= 7$

3. (2) 20

$$\begin{array}{rl} 60 & \text{(Thursday)} \\ -\ 40 & \text{(Wednesday)} \\ \hline 20 & \end{array}$$

4. (3) 30(9.95) + 4(4.00)

5. (5) not enough information is given
Although you know the square footage of the quilt, you do not know the measurements of each felt square.

6. (3) 12
Round off the 25 minutes to $\frac{1}{2}$ hour.
6 hours $\div \frac{1}{2}$ hour per exam = 12 exams

7. (3) $(2 \times 12) + 5$

8. (3) 5
$2{,}400 \div 500 = 4.8 \approx 5$ trips — Round up to find the number of trips Tim will need to make.

9. (4) 1,200

$$\frac{\text{envelopes}}{\text{hours}} = \frac{960}{4} = \frac{x}{5}$$

$960 \times 5 = 4 \times x$
$4{,}800 = 4 \times x$
$1{,}200 = x$

10. (2) $4.44

$ .49	$ .75	$1.50 coffee
× 6	× 2	+ 2.94 donuts
2.94	1.50	$4.44

11. (4) Southwest
The two bars are the closest.

12. (4) 35,000
40,000 cases cola
− 5,000 cases lemon seltzer
35,000 cases

13. (3) 15,000
$30{,}000 \times \frac{1}{2} = 15{,}000$

14. (5) not enough information is given
You need to know how *far* above cola and seltzer sales diet beverage sales rose.

15. (1) 5
9 ft × 5 ft = 45 sq ft
3 ft × 3 ft = 9 ft in 1 sq yd
45 sq yd ÷ 9 ft = 5 sq yd

16. (3) 880
$C = \pi d$
$5{,}280 = 3 \times d$
$5{,}280 \div 3 = d$
$1{,}760 = d$

$r = \frac{1}{2}d$
$r = \frac{1}{2} \times 1{,}760$
$r = 880$

17. (1) B, D, A, C
Convert all measurements to ounces:

A =	17 oz
B = $1\frac{1}{2}$ lb =	24 oz
C = 1 lb =	16 oz
D =	23 oz

Put in order from heaviest to lightest: B, D, A, C.

18. (3) $\frac{42 + 16}{2}$

19. (1) her gross pay for the 38 hours worked
To find Grace Anne's rate of overtime pay, you could subtract her regular pay (30 × $5.50) from her gross pay and divide that amount by the number of overtime hours she worked (8).

20. (2) 107
15 + 80 + 20 = 115 ounces
115 − 8 = 107 ounces

21. (3) 135
$P = 4s$
$540 = 4s$
$540 \div 4 = s$
$135 = s$

22. (5) not enough information is given
You need to know the number of guests to figure out how much the buffet will cost.

23. (3) $188.20
$19.95
8.00
12.50
12.50
$52.95

$52.95 × 4 trays of each = $211.80

$400.00 budget
− 211.80
$188.20

24. (3) 1,500

$75 per hour	75 guests	$1,275
× 3 hours	× $17	+ 225
$225 for hall rental	$1,275	$1,500

25. (3) $69.60
96 guests
× $\frac{1}{2}$ quart per person
48 quarts

$1.45 per quart
× 48 quarts
$69.60

26. (3) 10
$\frac{\text{inches long}}{\text{inches wide}} = \frac{7}{5} = \frac{14}{x}$
$7 \times x = 14 \times 5$
$7 \times x = 70$
$x = 70 \div 7$
$x = 10$

27. (4) 24.89
$9.95
5.95
+8.99
24.89

The information about the sundress and the change is unnecessary information.

28. (5) the amount of time it takes to hem one pair of pants

To find the amount the seamstress earns, you could multiply the number of pants (25) by the amount of time she spends on each pair, then multiply this total by her pay per hour ($8).

29. (4) 30

There are 30 minutes in $\frac{1}{2}$ hour.

30. (2) $2.00
$2 + 2\frac{1}{2} + 2\frac{1}{2} + 1\frac{1}{2} + 1 + \frac{1}{2} = 10$ hours
$95 ≈ $100
$100 ÷ 5 days = $20
$20 ÷ 10 hours = $2.00

Answer Key

Chapter 1: Looking at Word Problems

Exercise 1
The Numbers in Your Life
Page 3

Answers will vary.

Exercise 2
Reading Word Problems
Page 4

Answers may be similar to these.

1. The Crate Company hired some men for its warehouse. It also hired some women. How many people were hired in all?
2. A group of some committee members wanted to break down into smaller groups. They wanted a certain number of groups. How many people were in each group?
3. Over a period of time, a nurse received some calls. She spent an average of several minutes on each call. How much time in all did she spend on the telephone?
4. For the past several games batter José Alvarez has had a different number of hits each day. What is his average number of hits per game?
5. A tailor uses up a number of spools of thread each week. Each spool costs him a certain amount of money. How much does the tailor spend on thread each week?
6. Some mail carriers each worked a certain number of hours on Tuesday. If each mail carrier earns a certain amount of money per hour, how much was paid out on Tuesday?
7. When Todd works overtime, he earns several dollars per hour. His regular wage is a few dollars less. How much more does Todd earn per hour of overtime?
8. A woman started a trip at a certain speed, and she drove for a few hours at this rate. She sped up a bit and drove for another length of time. How many miles in all did she drive?

Exercise 3
Multiple-Choice Problems
Page 7

Your choices should include these correct answers.

1. **\$82** \$127 − 10 − 10 − 25 = 82
2. **15** (20 + 14 + 11) ÷ 3 = 15
3. **3** $4\frac{1}{2} - 1\frac{1}{2} = 3$
4. **\$535** \$500 × .07 = 35   35 + \$500 = \$535
5. **30** 10 + 10 + 5 + 5 = 30

Exercise 4
Multi-Step Word Problems
Page 8

1. **Step 1:** 120 workers − 36 sick = 84 workers
   Step 2: 84 workers + 20 temporary = **104 workers** (circled)
2. **Step 1:** \$ .72 + .04 = \$ .76
   Step 2: \$10.00 − .76 = **\$ 9.24** (circled)
3. **Step 1:** \$ .75 × 2 ways = \$1.50
   Step 2: \$ 1.50 × 14 days = **\$21.00** (circled)
4. **Step 1:** $8\frac{1}{2}$ hours × 5 days = 42.50 hours
   Step 2: 42.50 hours × 4 weeks = **170.00 hours** (circled)
5. **Step 1:** \$.48 × $1\frac{1}{2}$ lbs: .24 +.48 = \$.72
   Step 2: \$1.34 × $3\frac{1}{2}$: .67 +4.02 = \$4.69
   Step 3: \$ .72 + \$4.69 = **\$5.41** (circled)
6. **Step 1:** 20 feet × 40 feet = 800 square feet
   Step 2: 800 ÷ 100 sq ft = 8
   Step 3: 8 × 2 = **16 scoops** (circled)

Exercise 5
Word Problems with Visuals
Page 12

Part One

1. (4) 8,000,000

$$\begin{array}{r} 492 \text{ million} \\ -\ 484 \text{ million} \\ \hline 8 \text{ million} \end{array}$$

2. (5) 225,000,000

$$\begin{array}{r} 497 \text{ million} \\ -\ 272 \text{ million} \\ \hline 225 \text{ million} \end{array}$$

3. (2) $\frac{1}{2}$

$$\frac{226}{460} = \text{approximately } \frac{1}{2}$$

4. (4) 50%

$$\begin{array}{r} 429 \\ -\ 286 \\ \hline 143 \end{array} \qquad \frac{143}{286} = \frac{1}{2} = 50\%$$

Part Two

1. (4) 1,000

4,000 – 3,000 = 1,000
(rejected) (good)

2. (1) 17,000

3,000 + 6,000 + 8,000 = 17,000
(Jan) (Feb) (Mar)

3. (3) 100%

Jan → 3,000
Feb → 6,000 (increase of 3,000)
3,000 is 100% of 3,000

Chapter 2: The Five-Step Process

Exercise 1
Understand the Question
Page 16

Part One

Answers may be similar to these.

1. **Find:** the total money collected by all the Boy Scouts
2. **Find:** the number of exams each assistant will correct
3. **Find:** how much Marlene earned during her shift, including wages and tips
4. **Find:** how much money the salon receives from the hairdresser's earnings

Part Two

Questions will vary. Examples are given below.

1. How many hours did Lin study in all?
2. How many members voted against the budget cuts?
3. How many more miles did Sean drive than Jen?
4. How many square yards is the lot?

Exercise 2
Find the Information
Page 18

1. 32 ounces, 8 ounces, 16 ounces, 16 ounces, 32 ounces
2. 2 at $1.75, $.75, $.95, 2 at $.75
3. 211.4 miles, 12.1 gallons
4. 6 ounces of chocolate for each $1\frac{1}{2}$ cups sugar, 36 ounces chocolate
5. $2\frac{1}{2}$ feet, 4 inches

Exercise 3
Make a Plan
Page 21

1. **Divide**: 120 ÷ 30
2. **Add**: 148 + 320
3. **Subtract**: 33,004.1 – 32,908.4
4. **Multiply:** 2.2 × 110
5. **Add:** $3 + 3\frac{1}{2} + 2$

Exercise 4
Solve the Problem
Page 23

1. 4,439
2. 113
3. 33,354
4. 9,054
5. $25\frac{3}{4}$
6. 0.76
7. 40.5
8. 203
9. 64%
10. 220

If you got any of these wrong, you need to review your computation skills in that area.

Exercise 5
Check Your Answer
Page 25

1. a) 181 pounds

2. b) 2 hours

3. b) 288 fliers

4. b) 3 pints

5. a) 13 miles

6. b) 95 pounds

7. a) 12 questions

8. b) 24 inches

Chapter 3: Understanding the Question

Exercise 1
Read Carefully
Page 27

Part One

1. a. 25 miles
$$\begin{array}{r} 70 \\ -\ 45 \\ \hline 25 \end{array}$$

b. 210 miles
$$\begin{array}{r} 70 \\ 45 \\ +\ 95 \\ \hline 210 \end{array}$$

c. 70 miles
$$3\,\overline{)\,210}\quad = 70$$

d. 40 miles
$$\begin{array}{l} 250 \\ -\ 210 \text{ (altogether)} \\ \hline 40 \end{array}$$

2. a. 600 yards
$$\begin{array}{r} 75 \\ 75 \\ 225 \\ +\ 225 \\ \hline 600 \end{array}$$

b. 16,875 sq yd
$$\begin{array}{r} 225 \\ \times\ 75 \\ \hline 16{,}875 \end{array}$$

c. 3 times
$$75\,\overline{)\,225}\quad = 3$$

d. $21,093.75
$$\begin{array}{r} 16{,}875 \\ \times\ 1.25 \\ \hline 21{,}093.75 \end{array}$$

Part Two

Here are some possible answers. Yours may be different.

1. How much do the Phillipses earn together each month?
$990 + $1,050 = $2,040

How much more does Mrs. Phillips earn than her husband?
$1,050 – $990 = $60

2. How much was the meal altogether?
$5.50 + $7.00 + $4.25 + $2.00 = $18.75

If they split the bill evenly 3 ways, how much will each pay?
$18.75 ÷ 3 = $6.25

3. How much did Sue make per hour on average?
$160 ÷ 4 = $40

How much did Joe lose per hour on average?
$200 ÷ 4 = $50

Exercise 2
Some Tricky Questions
Page 31

1. a. (1) 1

Step 1:
$$\begin{array}{r} 15 \text{ minutes} \\ \times\ 4 \phantom{\text{ minutes}} \\ \hline 60 \text{ minutes} \end{array}$$

Step 2:
$$60\,\overline{)\,60 \text{ minutes}}\quad = 1 \text{ hour}$$
↑ minutes in an hour

b. (5) 60
$$\begin{array}{r} 15 \text{ minutes} \\ \times\ 4 \phantom{\text{ minutes}} \\ \hline 60 \text{ minutes} \end{array}$$

2. a. (3) 9

Step 1:
$$\begin{array}{r} 1{,}500 \\ 500 \\ +\ 2{,}500 \\ \hline 4{,}500 \end{array}$$

Step 2:
$$500\,\overline{)\,4{,}500 \text{ pounds}}\quad = 9 \text{ loads}$$
↑ pounds per load

b. (5) 4,500
$$\begin{array}{r} 1{,}500 \\ 500 \\ +\ 2{,}500 \\ \hline 4{,}500 \text{ pounds} \end{array}$$

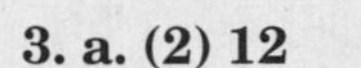

3. a. (2) 12

Step 1:

$$\begin{array}{r} 25.5 \\ \times \quad 4 \\ \hline 102.0 \text{ feet} \end{array}$$

Step 2:

$$8.5 \overline{)\,102} \quad = 12$$

b. (5) $479.40

$$\begin{array}{r} \$4.70 \\ \times\ 102 \text{ feet} \\ \hline \$\ 479.40 \end{array}$$

4. a. (5) 2,200

Step 1:

$$\begin{array}{r} 440 \text{ yards} \\ \times\ 4 \\ \hline 1{,}760 \text{ yards} \end{array}$$

Step 2:

$$\begin{array}{r} 1{,}760 \text{ yards} \\ +\ 440 \text{ yards} \\ \hline 2{,}200 \text{ yards} \end{array}$$

b. (1) $1\frac{1}{4}$ $1 \text{ mile} + \frac{1}{4} \text{ mile} = 1\frac{1}{4} \text{ miles}$

Exercise 3
Working with "Set-Up Questions"
Page 34

1. (3) $240 + 700$
2. (2) 3000×3
3. (5) $64 - 59$
4. (2) $70 \div 8$
5. (3) $7.10 \times .06$
6. (5) 9×4

Exercise 4
More on "Set-Up Problems"
Page 38

Part One

1. $(60 \times 3) + (65 \times 2)$

hours ↓ (3), hours ↓ (2); mph ↑ (60), mph ↑ (65)

2. $\dfrac{12 + 15}{3}$

girls ↓ (12), boys ↓ (15); groups ↑ (3)

3. $2606 - (1194 + 457)$

miles on second day ↓ (1194), miles on first day ↓ (457)

4. $(4 \times 1.29) + (4 \times .89)$ OR $4\,(1.29 + .89)$

cost ↓ (1.29), cost ↓ (.89); cans ↑ (4)

5. $2\,(.35 \times \$3.99)$ OR $.35\,(2 \times \$3.99)$

percent off ↓ (.35); number of shirts ↑ (2), original price ↑ ($3.99)

6. $\$.11(4 + 9 + 16)$

cost per foot ↑ ($.11), sides ↑ (4 + 9 + 16)

Part Two

1. (3) $\dfrac{13 + 11 + 15 + 10}{4}$

miles (13 + 11 + 15 + 10); 4 ← days

2. (5) $3{,}300 - (2 \times 390)$

maximum milligrams ↓ (3,300), milligrams per muffin ↓ (390); number of muffins ↑ (2)

3. (5) $1{,}000(8 - 2)$

July ↓ (8), October ↓ (2); thousands of dollars ↑ (1,000)

4. (2) $(890 - 50) + 110$

balance on Thursday ↓ (890), deposit ↓ (110); withdrawal ↑ (50)

5. (3) $\frac{1}{3}(800{,}000{,}000) + 800{,}000{,}000$

U.S. population is $\frac{1}{3}$ India's ↓ ($\frac{1}{3}$(800,000,000)), India's population ↓ (800,000,000)

6. (2) $\frac{1}{3}(10 \times 24)$

amount to be covered ↓ ($\frac{1}{3}$); width ↑ (10), length ↑ (24)

Mixed Review
Page 41

1. (4) $(2 \times 100) + (2 \times 85)$ — mph ↓ above 100 and 85; hours ↑ below each 2

2. (2) $(1{,}043 + 357) + 34$ — 1987 law officers ↓ (1,043); number down from 1986 ↓ (34); 1987 civilians ↑ (357)

3. **(3) 12**

   Step 1: 4 + 8 + 8 + 12 + 12 + 4 = 48 hours
   Step 2: 48 ÷ 4-hour shifts = 12 shifts

4. **(2) 6**

   Step 1: 6.2 + 4.2 + 6.2 + 5.4 + 6.2 + 7.8 = 36
   Step 2: 36 ÷ 6 = 6

5. **(4) $66\frac{2}{3}\%$**

   $$\frac{\text{people on tour bus}}{\text{total people}} = \text{percent on tour bus}$$

   $$\frac{8{,}000}{12{,}000} = \frac{8}{12} = \frac{2}{3} \qquad \frac{2}{3} = 66\tfrac{2}{3}\%$$

6. (3) $(2 \times 220) + .07(2 \times 220)$ — number of suits ↓ (each 2); cost ↑ (220), tax ↑ (.07), cost ↑ (220)

7. **(2) $45,750**

   Step 1: .06 × $62,500 = $3,750 commission
   Step 2: $62,500 – ($3,750 + $13,000) = $45,750

8. **(3) 75**

   Step 1:
   25 feet
   × 12 inches per foot
   300 inches

   Step 2:
   $300 \div 4 = 75$

9. **(3) 6**

   Step 1: 4 + 2 + 6 = 12 cups
   Step 2: 12 ÷ 2 cups per pint = 6 pints

10. **(2) 5**

    Step 1:
    5 × 15 = 75
    10 × 15 = 150
    15 × 15 = 225
    75 + 150 + 225 = 450 sq ft

    Step 2:
    450 ÷ 9 square feet in a square yard = 50 sq yd

    Step 3:
    50 ÷ 10 = 5 jars

Chapter 4: Finding the Information

Exercise 1
Looking at Labels
Page 45

1. 11 acres × 120 trees = **1,320 trees**
2. 4 Chevys + 10 Fords + 10 Chryslers + 9 Toyotas = **33 cars**
3. 900 miles ÷ $2\frac{1}{2}$ hours = **360 miles per hour**
4. 10 meters × 39 inches = **390 inches**
5. 17 ft × 20 ft = **340 sq ft**

Exercise 2
Finding "Hidden" Information
Page 48

1. **$880**

   $ 440 room and board
   × 2 *(twice)*
   $ 880

2. **2 hours**

   Step 1: 15 + 15 = 30 blocks
   Step 2: 30 ÷ 15 blocks per hour = 2 hours

3. **38 people**

   3 + 8 + 17 + 10 = 38 people

4. **30 pounds**

   Step 1: 5 bags × 8 pounds = 40 pounds
   Step 2: 40 – 10 = 30 pounds

5. **$13.75**

   Step 1: 3 + 10 + 12 = 25 sheets
   Step 2: 25 × $.55 = $13.75

6. $950

Step 1:

$$\begin{array}{r} 118 \\ +\ 72 \\ \hline 190 \end{array}$$

Step 2:

$$\begin{array}{r} 190 \\ \times\ \$5 \\ \hline \$950 \end{array}$$

7. 6,400,000 dairy cows

64% of 10,000,000 = .64 × 10,000,000 = 6,400,000

8. 8 hours

Step 1:

8:00 until 6:00 = 10 hours work
+ 6 hours sleep
16 hours

Step 2:

24 hours in a day
− 16 hours
8 hours left

9. 10 poles

Step 1: 10 yards = 30 feet

Step 2: 30 feet ÷ 3 = 10 poles

10. $540

Step 1:

Area of triangle = $\frac{1}{2}bh$

Area of patio = $\frac{1}{2}(15 \times 12)$
= $\frac{1}{2}(180)$
= 90 square feet

Step 2: 90 × $6 = $540

Exercise 3
Extra Information
Page 51

Part One

1. **Question:** How much is the car with a sunroof and rustproofing?
   Necessary Information: $7,995; $350; $200

2. **Question:** By how much had the driver exceeded the speed limit the first time?
   Necessary Information: 82 mph; 55 mph

3. **Question:** How much money was collected for adult tickets?
   Necessary Information: twice 100 tickets (2 × 100); $5.50

4. **Question:** How much did the dairy bar take in for large cones?
   Necessary Information: 22,500 cones: 35% large cones; $.55

Part Two

1. **33 miles per gallon**
   Extra Information: 7 hours
   Solution: 396 miles ÷ 12 gallons = 33 miles per gallon

2. **$49**
   Extra Information: $12.06 and $12.06 in wages
   Solution: ($51 + $32 + $64) ÷ 3 = $49

3. **8 grams**
   Extra Information: all cholesterol; fat—butter, 11; margarine, 11
   Solution: 11 − 3 = 8 grams

4. **1 movie**
   Extra Information: 3 movies on TV; twice as many
   Solution: $\frac{1}{2} \times 2$ = 1 movie

5. **.3%**
   Extra Information: 4.8%; 4.6%
   Solution: 10.3% − 10.0% = .3%

Exercise 4
Not Enough Information
Page 56

Part One

1. **Missing Information:** number of students on Tuesdays and Thursdays

2. **Missing Information:** cost of utilities at Mr. Santos's apartment

3. **Missing Information:** amount of tax *or* percent of tax on the sheet music

4. **Missing Information:** number of hours Mrs. Sobala spends at work *or* what time Mrs. Sobala leaves work

5. **Missing Information:** number of residents who voted

Part Two

1. **(5) not enough information is given**
   You need to know the cost of a gallon of gas.

2. (1) $1.95

Step 1:	Step 2:
\$ 2.70	\$ 5.75
+ 1.10	– 3.80
\$ 3.80	\$ 1.95

3. (2) $11.31

Step 1: $.06 \times \$36.50 = 2.19$

Step 2:	Step 3:
\$ 36.50	\$ 50.00
+ 2.19	– 38.69
\$ 38.69	\$ 11.31

4. (5) not enough information is given
You need to know the amount of time he spent refueling.

5. (5) not enough information is given
The graph shows people who were laid off. You need to know how many people *were employed*.

Exercise 5
What More Do You Need to Know?
Page 58

1. **(3)** the number of books in a crate
2. **(4)** the amount of money she gave the cashier
3. **(1)** the amount of gas his tank holds
4. **(3)** the percent of people who responded to the questionnaire
5. **(5)** the width of the floor

Exercise 6
Working with "Item Sets"
Page 62

A. 1. Necessary Information
8 lines, 4 lines, 4 workers per line
Answer:
Step 1: 8 lines + 4 lines = 12 lines
Step 2: 12 lines × 4 workers = **48 workers**

2. Necessary Information:
8 lines of 4 workers at $6 for 8 hours
4 lines of 4 workers at $9 for 8 hours
Answer:
Step 1:
day shift = 8 lines × 4 workers = 32 workers
32 workers × 8 hours × \$6 = \$1,536
night shift = 4 lines × 4 workers = 16 workers
16 workers × 8 hours × \$9 = \$1,152
Step 2: \$1,536 + \$1,152 = **\$2,688**

3. Necessary Information: 80 boxes per hour, 8 hours, 8 lines
Answer: 80 × 8 × 8 = **5,120 boxes**

B. 1. Necessary Information:
length = 20 feet; width = 12 feet
$\frac{3}{4}$ grass
Answer: A = l × w
Step 1: A = 20 feet × 12 feet
A = 240 sq ft
Step 2: $240 \times \frac{3}{4}$ = **180 sq feet**

2. Necessary Information:
240 sq ft; $\frac{3}{4}$; 4 bricks per square foot
Answer:
Step 1: 240 sq ft – 180 sq ft = 60 sq ft
Step 2: 60 sq ft × 4 bricks = **240 bricks**

3. Necessary Information: 240 bricks (from problem 2); \$.20 per brick; \$20 bag of mortar; \$320
Answer:
Step 1: 240 × \$.20 = \$48.00
\$48 + \$20 = \$68.00
Step 2: \$320 – 68 = **\$252**

Exercise 7
Making Charts to Solve Item Sets
Page 65

Part One

	1st Project	2nd Project
Sasha's pay at \$9 per hour	\$9 × 100 = \$900	8 hours × 20 days = 160 hours 160 × \$9 = \$1,440
Computer operator's pay at \$7 per hour	\$7 × 10 = \$70	0
Cost of materials	\$124	\$280

1. **(3) \$156** \$280 – \$124 = \$156
2. **(5) \$1,094** \$900 + \$70 + \$124 = \$1,094
3. **(3) \$1720** \$1,440 + \$280 = \$1,720

Part Two

cost of carrots	.69/lb or $3 for 5 lbs
restaurant purchases	20 lbs carrots 25 lbs flour
discount rates	15% except vegetables

1. (2) $1.80

Step 1:
20 lbs ÷ 5 lbs = 4 bags
$3 × 4 bags = $12.00

Step 2:
20 lbs × $.69 = $13.80

Step 3:
$13.80 – $12.00 = $1.80

2. (3) $6.63

Step 1:
3 lbs × $2.60 = $7.80

Step 2:
$7.80 × 15% discount = $1.17

Step 3:
$7.80 – $1.17 = $6.63

3. (5) not enough information is given
You need to know how much the store charges for flour.

Part Three

Yards needed	2	$\frac{3}{4}$	$\frac{3}{4}$	$3\frac{1}{2}$
cost per yard	$4.00/yd	$4.00/yd	$4.00/yd	$4.00/yd
tax	5%	5%	5%	5%

1. (3) $3\frac{1}{2}$ **Step 1:** $2 + \frac{3}{4} + \frac{3}{4} = 3\frac{1}{2}$

2. (4) 2 **Step 1:** $5\frac{1}{2} - 3\frac{1}{2} = 2$

3. (4) $23.10 **Step 1:** $5\frac{1}{2} \times \$4.00 = \$22.00$
Step 2: $22.00 + 5%($22.00) = $23.10

Mixed Review

Page 68

1. (2) 80

$$\begin{array}{r} 420 \\ -\ 340 \\ \hline 80 \end{array}$$

2. (4) 24

Step 1:
12 + 27 + 33 = 72

Step 2:
$$3 \overline{)72} \quad = 24$$

3. (1) 3.2

$$\begin{array}{rl} 15.3 & \text{seconds} \\ -\ 12.1 & \text{seconds} \\ \hline 3.2 & \text{seconds} \end{array}$$

4. (3) $1,732.50

Step 1:
$\frac{1}{2} \times \$3.50 = \$1.75$

Step 2:
435 + 224 + 331 = 990 tickets

Step 3:
$$\begin{array}{r} 990 \\ \times\ \ 1.75 \\ \hline \$1{,}732.50 \end{array}$$

5. (5) $\frac{11}{20}$

Step 1:
$$\begin{array}{rcl} \frac{1}{4} & = & \frac{5}{20} \\ \frac{1}{5} & = & \frac{4}{20} \\ +\ \frac{2}{5} & = & \frac{8}{20} \\ \hline & & \frac{17}{20} \end{array}$$
Monday, Friday, and Saturday

Step 2:
$$\begin{array}{rcl} 1 & = & \frac{20}{20} \\ -\ \frac{17}{20} & = & \frac{17}{20} \\ \hline & & \frac{3}{20} \end{array}$$
Sunday

Step 3:
$\frac{8}{20}$ Saturday
$+\ \frac{3}{20}$ Sunday
$\frac{11}{20}$ on the weekend

6. (4) $\frac{15 + 75 + 55}{60}$

(15 + 75 + 55: total minutes; 60: minutes per hour)

7. (1) $20

Step 1:
Bert's = $305 + 75 = 380
Tech TV = $400

Step 2:
$ 400
– 380
$ 20

8. (3) $305 – $150
↑ Bert's ↑ repair cost

Chapter 5: Making a Plan

Exercise 1
Choosing the Operation
Page 72

1. **Operation:** ÷ Answer: 5
2. **Operation:** – Answer: 220
3. **Operation:** + Answer: 161
4. **Operation:** × Answer: 48
5. **Operation:** – Answer: 6
6. **Operation:** – Answer: $4.15
7. **Operation:** × Answer: $4.00
8. **Operation:** ÷ Answer: 5
9. **Operation:** – Answer: 9 inches
10. **Operation:** ÷ Answer: 26

Exercise 2
Equations
Page 75

Part One

1. b) 5 – 2 = 3 hours

2. a) 220 – 117 = 103 miles

3. b) 140 + 30 = 170 cans

4. a) $.70 × 35 = $24.50

5. a) $6 \div \frac{1}{3}$ = 18 stripes

Part Two

Although you may have used different words, your plans should be similar to these.

1. **Plan:** I should multiply the number of days by the hours per day.
Equation: 6 days × 8 hours = **48 hours**
2. **Plan:** I should add the minutes it takes to wash to the minutes it takes to dry.
Equation: 17 minutes + 45 minutes = **62 minutes**
3. **Plan:** I should subtract the miles driven from the total miles.
Equation: 300 miles – 76 miles = **224 miles**
4. **Plan:** I should multiply the number of shirts by the cost per shirt.
Equation: 6 × $1.00 = **$6.00**
5. **Plan:** I should divide the total people by the number of people per group.
Equation: 225 people ÷ 15 people per group = **15 groups**
6. **Plan:** I should add the money earned by one family to the money earned by the other family.
Equation: $87 + $112 = **$199**
7. **Plan:** I should divide the length of the path by the distance between the trees.
Equation: 152 ÷ 8 = **19 trees**
8. **Plan:** I should multiply the total the sandbox holds by the fraction that is filled.
Equation: $3 \times \frac{1}{2} = 1\frac{1}{2}$ **cubic feet**
9. **Plan:** I should multiply the percent in the vocational program by the total in the school.
Equation: 14% × 800 students = 112 students
10. **Plan:** I should multiply the length of one side by the number of sides.
Equation: 10 inches × 4 sides = **40 inches**

Exercise 3
Equations with Two Operations
Page 79

Part One

1. b) $(18 \div 3) \div 12 = \frac{1}{2}$

2. a) (36 ÷ 4) + 3 = 12

3. b) ($64.89 – $14.01) ÷ .06 = 848

4. a) (1.75 + 2.8) × $1,244 = $5,660.20

5. a) (24 × 20) ÷ 2 = 240

Part Two

1. **Equation:** (36 – 14) ÷ 22 = **1 cupcake**
2. **Equation:** (15 × 200) ÷ 400 = **7.5 minutes**
3. **Equation:** $19.80 ÷ (3.5 + 2) = **$3.60 per pound**

4. Equation: $\$7.10\ (8\frac{1}{2} + 4\frac{1}{2} + 6\frac{1}{2}) =$ **$138.45**

5. Equation: $(5\frac{1}{2} + 8\frac{1}{2}) \div 2 =$ **7 miles**

6. Equation: $.20\ (4 + 8 + 5 + 3 + 5) =$ **5 pounds**

7. Equation: $(8 \times 7) \div 28 =$ **2 bags**

Exercise 4
Writing Equations for Word Problems
Page 84

Part One

1. x = additional hours to be worked
$28 + x = 40$ OR $40 - 28 = x$

2. x = minutes Sally and Doug talked
$.90 \times x = \$5.40$ OR $\$5.40 \div .90 = x$

3. x = cost of a pair of pants
$3 \times x = \$44.99$ OR $\$44.99 \div 3 = x$

4. x = width of room
$12 \times x = 120$ OR $120 \div 12 = x$

Part Two

You could have written **two** of the equations listed for each problem below.

1. x = hours worked per day
$35 \div 7 = x$
$35 \div x = 7$
$x \times 7 = 35$
$7 \times x = 35$

2. x = people outside city
$29 - x = 7$
$29 - 7 = x$
$7 + x = 29$
$x + 7 = 29$

3. x = number of daisies in a bunch
$x \times 80 = 720$
$80 \times x = 720$
$720 \div 80 = x$
$720 \div x = 80$

4. x = $ to spend for each boy
$x \times 7 = 21$
$7 \times x = 21$
$21 \div x = 7$
$21 \div 7 = x$

Exercise 5
Solving a Word Problem Equation
Page 89

Part One

1. $37 + x = 91$
$x = 91 - 37$
$\mathbf{x = 54}$

2. $x \div 13 = 11$
$x = 11 \times 13$
$\mathbf{x = 143}$

3. $144 \div 4 = x$
$\mathbf{36 = x}$

4. $x - 194 = 19$
$x = 19 + 194$
$\mathbf{x = 213}$

5. $80 - x = 14$
$80 - 14 = x$
$\mathbf{66 = x}$

6. $14 \times x = 42$
$x = 42 \div 14$
$\mathbf{x = 3}$

7. $144 \div x = 12$
$144 \div 12 = x$
$\mathbf{12 = x}$

8. $27 + x = 120$
$x = 120 - 27$
$\mathbf{x = 93}$

9. $x \div 12 = 9$
$x = 9 \times 12$
$\mathbf{x = 108}$

10. $105 \div x = 7$
$105 \div 7 = x$
$\mathbf{15 = x}$

Part Two

1. $96 \div x = 32$
$96 \div 32 = x$
$3 = x$

Answer: 3 children

2. $x + 12 = 60$
$x = 60 - 12$
$x = 48$

Answer: 48 pounds

3. $x \times 15 = 225$
$x = 225 \div 15$
$x = 15$

Answer: 15 committees

4. $x - 145 = 1{,}004$
$x = 1{,}004 + 145$
$x = 1{,}149$

Answer: $1,149

5. $360 \div 40 = x$
$9 = x$

Answer: 9 pans

Exercise 6
Writing Proportions for Word Problems
Page 91

Part One

1. $\frac{miles}{hours}$ $\frac{900}{16} = \frac{225}{x}$

2. $\frac{inches}{miles}$ $\frac{1}{300} = \frac{3}{x}$

3. $\frac{chocolate}{sugar}$ $\frac{7}{3} = \frac{21}{x}$

4. $\frac{ounces}{cost}$ $\frac{20}{4.39} = \frac{x}{5.28}$

Part Two

1. $\frac{good\ strands}{bad\ strands}$ $\frac{15}{2} = \frac{x}{1{,}000}$ OR $\frac{bad}{good}$ $\frac{2}{15} = \frac{1{,}000}{x}$

2. $\frac{miles}{gallons}$ $\frac{352}{11} = \frac{x}{4}$ OR $\frac{gallons}{miles}$ $\frac{11}{352} = \frac{4}{x}$

3. $\frac{blue}{white}$ $\frac{2}{3} = \frac{5\frac{1}{2}}{x}$ OR $\frac{white}{blue}$ $\frac{3}{2} = \frac{x}{5\frac{1}{2}}$

4. $\frac{liters}{minutes}$ $\frac{4{,}800}{50} = \frac{x}{90}$ OR $\frac{minutes}{liters}$ $\frac{50}{4{,}800} = \frac{90}{x}$

5. $\frac{width}{length}$ $\frac{3}{5} = \frac{x}{20}$ OR $\frac{length}{width}$ $\frac{5}{3} = \frac{20}{x}$

Exercise 7
Solving Proportion Equations
Page 95

1. $\frac{x}{5} = \frac{4}{10}$
   $x \times 10 = 5 \times 4$
   $x \times 10 = 20$
   $x = 20 \div 10$
   $x = 2$

2. $\frac{8}{4} = \frac{60}{x}$
   $8 \times x = 60 \times 4$
   $8 \times x = 240$
   $x = 240 \div 8$
   $x = 30$

3. $\frac{x}{16} = \frac{200}{80}$
   $x \times 80 = 200 \times 16$
   $x \times 80 = 3{,}200$
   $x = 3{,}200 \div 80$
   $x = 40$

4. $\frac{9}{x} = \frac{6}{4}$
   $9 \times 4 = 6 \times x$
   $36 = 6 \times x$
   $36 \div 6 = x$
   $6 = x$

5. $\frac{13}{5} = \frac{x}{10}$
   $13 \times 10 = 5 \times x$
   $130 = 5 \times x$
   $130 \div 5 = x$
   $26 = x$

6. $\frac{8}{3} = \frac{24}{x}$
   $8 \times x = 24 \times 3$
   $8 \times x = 72$
   $x = 72 \div 8$
   $x = 9$

Part Two

You may have written the proportions in a different order (reversed top and bottom), but your answers should be the same.

1. $\frac{15}{2} = \frac{x}{1{,}000}$
   $15 \times 1{,}000 = 2 \times x$
   $15{,}000 = 2 \times x$
   $15{,}000 \div 2 = x$
   $\mathbf{7{,}500 = x}$

2. $\frac{352}{11} = \frac{x}{4}$
   $352 \times 4 = x \times 11$
   $1{,}408 = x \times 11$
   $1{,}408 \div 11 = x$
   $\mathbf{128 = x}$

3. $\frac{2}{3} = \frac{5\frac{1}{2}}{x}$
   $2 \times x = 5\frac{1}{2} \times 3$
   $2 \times x = 16\frac{1}{2}$
   $x = 16\frac{1}{2} \div 2$
   $\mathbf{x = 8\frac{1}{4}}$

4. $\frac{4{,}800}{50} = \frac{x}{90}$

$4{,}800 \times 90 = 50 \times x$

$432{,}000 = 50 \times x$

$432{,}000 \div 50 = x$

$\mathbf{8{,}640 = x}$

5. $\frac{3}{5} = \frac{x}{20}$

$3 \times 20 = x \times 5$

$60 = x \times 5$

$60 \div 5 = x$

$\mathbf{12 = x}$

Exercise 8
When Can You Use a Proportion?
Page 97

Part One

1. Yes

part-time	1	13
full-time	6	x

2. Yes

cost	\$15	x
diapers	40	100

3. Yes

yogurt	2	$1\frac{1}{2}$
fruit	5	x

4. Yes

people	1,600	x
percent	100%	70%

5. No

Part Two

1. 78 full-time employees

$1 \times x = 6 \times 13$

$x = 78$

2. \$37.50

$\$15 \times 100 = x \times 40$

$\$1{,}500 = x \times 40$

$\$1{,}500 \div 40 = x$

$\$37.50 = x$

3. $3\frac{3}{4}$ pounds

$2 \times x = 5 \times 1\frac{1}{2}$

$2 \times x = 7\frac{1}{2}$

$x = 7\frac{1}{2} \div 2$

$x = \frac{15}{2} \times \frac{1}{2} = \frac{15}{4} = 3\frac{3}{4}$

4. 1,120 people

$15\% \times 1600 = x$

$1{,}120 = x$

5. cannot use proportion

Exercise 9
Drawing a Picture
Page 101

Part One

1. $960 \div 24 =$ **40 petitions**

2. $8 - 2\frac{1}{2} =$ **$5\frac{1}{2}$ feet**

3. $99 \div 3 =$ **33 loaves**

4. $27 \times \frac{1}{2} =$ **$13\frac{1}{2}$ miles**

5. $34\frac{1}{2} - 31\frac{1}{4} =$ **$3\frac{1}{4}$ inches**

6. Step 1: $10 \times 5 \times 6 =$ **300 cartons**
Step 2: 300 cartons × 8 cans = **2,400 cans**

Part Two

Your drawings will probably look different from these. Use whatever works for you.

1.

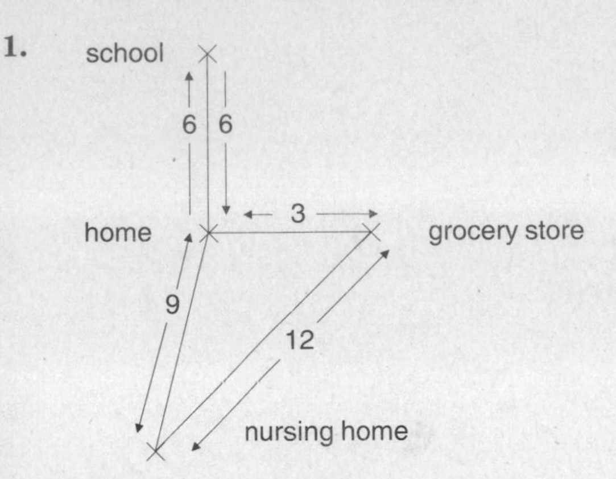

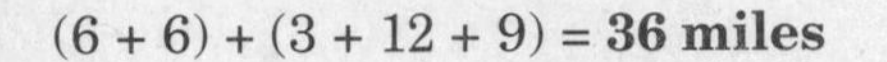

(6 + 6) + (3 + 12 + 9) = **36 miles**

2.

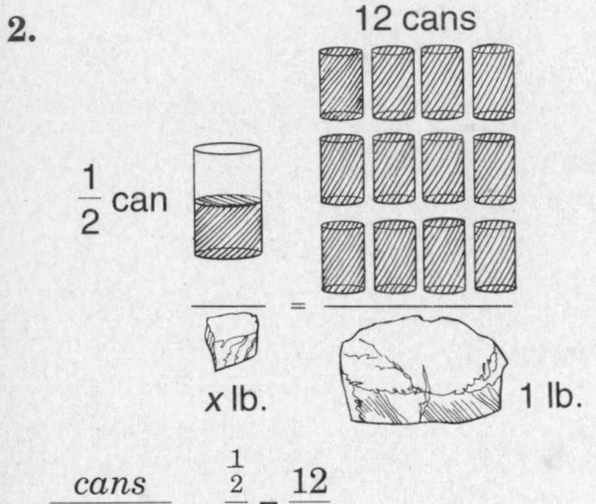

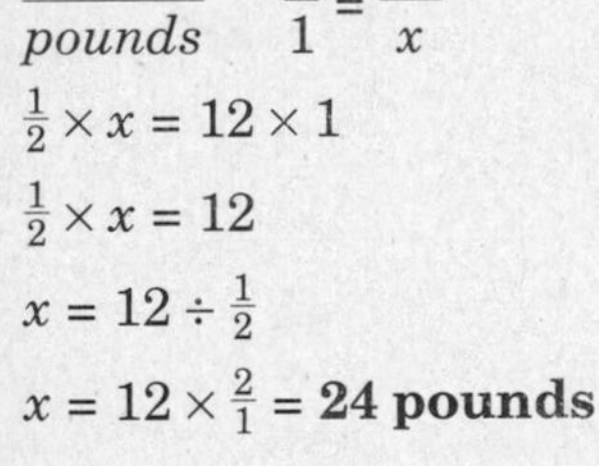

$\frac{\textit{cans}}{\textit{pounds}} \quad \frac{\frac{1}{2}}{1} = \frac{12}{x}$

$\frac{1}{2} \times x = 12 \times 1$

$\frac{1}{2} \times x = 12$

$x = 12 \div \frac{1}{2}$

$x = 12 \times \frac{2}{1} =$ **24 pounds**

3.

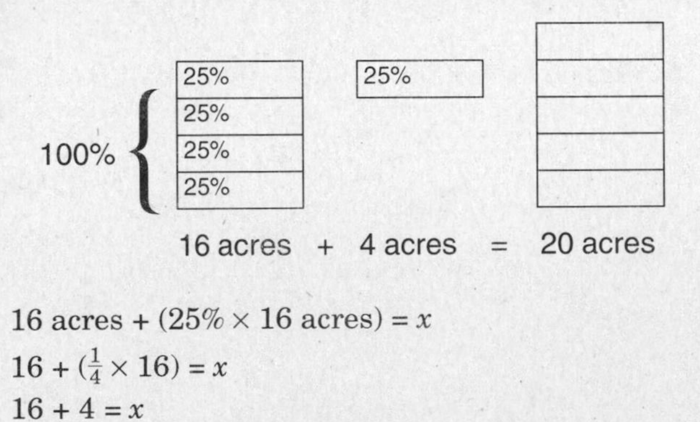

16 acres + (25% × 16 acres) = x

16 + ($\frac{1}{4}$ × 16) = x

16 + 4 = x

20 acres = x

4.

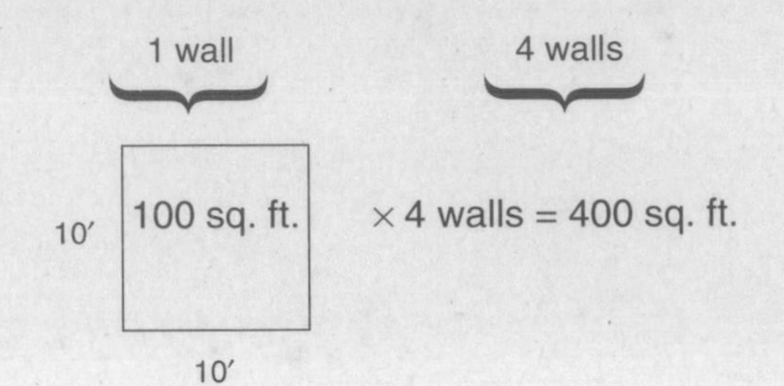

Step 1: 10 × 10 = **100 square feet**
Step 2: 4 × 100 = **400 square feet**

5.

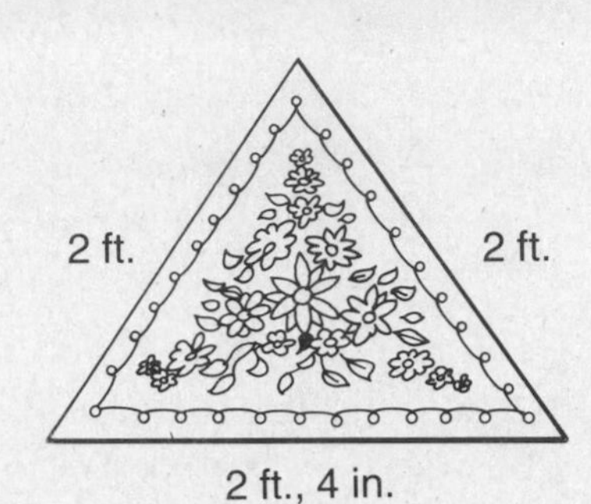

2 ft + 2 ft + 2 ft, 4 in = **6 ft, 4 in**

6.

2 × 85 = **170 yds**

Mixed Review
Page 104

1. (2) 10
6 gallons × 2 apartments = 12 gallons needed
12 gallons − 2 on hand = 10 gallons

2. (5) $\frac{\mathbf{400 + 250}}{\mathbf{50}}$

syrup ↓ antibiotic ↓
$\frac{\mathbf{400 + 250}}{\mathbf{50}}$
↑
size of jar

3. (3) 48

$\frac{\textit{formula}}{\textit{water}} \quad \frac{2 \text{ oz}}{6 \text{ oz}} = \frac{16}{x}$

$2 \times x = 16 \times 6$

$2 \times x = 96$

$x = 96 \div 2$

$x = 48$ ounces

4. (4) 75°
Step 1: 70 + 70 + 80 + 85 + 75 + 85 + 60 = 525
Step 2: 525 ÷ 7 = 75

5. (4) $3\frac{3}{4}$

$7\frac{1}{2} - 3\frac{3}{4} = 3\frac{3}{4}$ miles

6. (1) 14.85

30% × 49.5 = 14.85 hours

7. (3) $7,350

Step 1: $12,000 ÷ 2 = $6,000
Step 2: $13,350 – $6,000 = $7,350

8. (4) $480

Step 1:
Rachel's monthly payment: 40% × $750 = $300
Step 2:
monthly payment + utilities: $300 + $180 = $480

Chapter 6: Solving the Problem

Exercise 1
Keeping Organized
Page 108

1. a) 11 hits
Step 1: 12 + 11 + 10 = 33 hits (top 3 players)
Step 2: 33 ÷ 3 = 11 hits on average
b) Incorrect solution: 8 hits is the average of **all** players. The question asks for the average of the **top 3** players.

2. a) 16 hours
Step 1: $4 \times 1\frac{1}{2} = 6$ hours mowing
2 × 1 = 2 hours fixing drains
8 hours remodeling
Step 2: 6 + 2 + 8 = 16 hours
b) Incorrect solution: The student forgot to add in the hours spent on remodeling.

Exercise 2
Estimating the Answer
Page 110

Your estimates may be slightly different from what you see here.

1. a) Estimate: 211 ≈ 200 pounds
68 ≈ 70 pounds
41 ≈ 40 pounds
Step 1: 200 – (70 + 40) = x
Step 2: 200 – 110 = **90**
b) Answer choice: **(4) 102** (102 is closest to 90)
c) Exact solution:
Step 1: 211 – (68 + 41) = x
Step 2: 211 – 109 = **102 lbs**

2. a) Estimate: $1.48 ≈ $1.50
3.25 ≈ 3 lbs
$1.50 × 3 = **$4.50**
b) Answer choice: **(2) $4.81** (closest to $4.50)
c) Exact solution: $1.48 × 3.25 = **$4.81**

3. a) Estimate: 1.86 ≈ $2.00
10.5 ≈ 10 gallons
10 × 2 = **$20.00**
b) Answer choice: **(4) $19.53**
c) Exact solution: $1.86 × 10.5 = **$19.53**

4. a) Estimate: 403 ≈ 400 students
2,015 ≈ 2,000 students
400 ÷ 2,000 = .20 = 20%
b) Answer choice: **(3) 20**
c) Exact solution: 403 ÷ 2,015 = .20 = **20%**

Exercise 3
Estimating with Fractions
Page 113

Your estimates may be slightly different from the ones shown here. That doesn't matter—as long as your estimates help you decide what to do.

1. a) Estimate: $24\frac{1}{2} \approx 25$
$9.70 ≈ $10
25 × $10 = **$250**
b) Answer choice: **(2) $237.65**
c) Exact solution:

$$\begin{array}{r} 24.5 \\ \times\ \$9.70 \\ \hline \mathbf{\$\,237.65} \end{array}$$

2. a) Estimate: $24\frac{5}{8} \approx 25$
$10\frac{1}{4} \approx 10$
Estimate: 25 – 10 = **15**
b) Answer choice: **(3)** $\mathbf{14\frac{3}{8}}$
c) Exact solution: $24\frac{5}{8} - 10\frac{1}{4} =$
$24\frac{5}{8} - 10\frac{2}{8} = \mathbf{14\frac{3}{8}}$

3. a) Estimate: $3\frac{1}{2} \approx 4$
$2\frac{1}{4} \approx 2$
Estimate: 4 × 2 = **8**
b) Answer choice: **(3)** $\mathbf{7\frac{7}{8}}$
c) Exact solution: $3\frac{1}{2} \times 2\frac{1}{4} =$
$\frac{7}{2} \times \frac{9}{4} = \frac{63}{8} = \mathbf{7\frac{7}{8}}$

4. a) Estimate: $2\frac{1}{5} \approx 2$

$3\frac{1}{4} \approx 3$

$3\frac{3}{10} \approx 3$

$4\frac{7}{10} \approx 5$

$2 + 3 + 3 + 5 = \mathbf{13}$

b) Answer choice: **(1)** $\mathbf{13\frac{9}{20}}$

c) Exact solution: $2\frac{1}{5} + 3\frac{1}{4} + 3\frac{3}{10} + 4\frac{7}{10}$

$2\frac{4}{20} + 3\frac{5}{20} + 3\frac{6}{20} + 4\frac{14}{20} = \mathbf{13\frac{9}{20}}$

• • • • • • • •

Exercise 4
When an Estimate Is the Answer
Page 114

1. (2) 6
$23 \approx 24$

24 sales ÷ 4 sales = 6 times

2. (3) 8
1 hour = 60 minutes
60 ÷ 7 minutes = *about* 8

3. (1) 9
Step 1: January sales: about 450
July sales: about 50
Step 2: 450 ÷ 50 = 9

4. (5) $4,600
Step 1: January sales: about 450
February sales: about 400
March sales: about 300
ticket price: about $4.00
Step 2: 450 + 400 + 300 = 1,150
Step 3: 1,150 × $4.00 = $4,600

5. (4) 25%
Step 1: April sales: about 150
May sales: about 150
June sales: about 100
Step 2: Total = 150 + 150 + 100
Total = 400
Step 3: $\frac{\text{June}}{\text{Total}} = \frac{100}{400} = \frac{1}{4} = 25\%$

• • • • • • • •

Exercise 5
Writing Answers in Set-Up Format
Page 117

Your answers may include any of these.

1. a) (3 × 55) + (3 × 60)
b) 3 (55 + 60)
c) 3 (55) + 3 (60)

2. a) 32.80 – (2 × 3.00)
b) 32.80 – 6

3. a) $36 - \frac{1}{4}(36)$
b) $\frac{3}{4}(36)$
$(1\text{ hr} - \frac{1}{4}\text{ hr} = \frac{3}{4}\text{ hr})$
c) 36 – (36 ÷ 4)

4. a) 40 + 40 + 20 + 20
b) (2 × 40) + (2 × 20)
c) 2 (40 + 20)

• • • • • • • •

Exercise 6
Comparing and Ordering Numbers
Page 120

1. a) LABEL: dollars
b) ANSWER: **(4) D**

A. $4,800
B. $5,500
C. $4,600
D. $4,500
E. $5,400

2. a) LABEL: % in favor
b) ANSWER: **(1) B, A, D, C, E**

Town A. 62%
Town B. $66\frac{2}{3}\%$
Town C. 46%
Town D. 52%
Town E. $33\frac{1}{3}\%$

3. a) LABEL: feet (or inches)
b) ANSWER: **(2) D, E, C, B, A**

Board A. $2\frac{1}{2}$ feet (30 inches)
Board B. 3 feet (36 inches)
Board C. $3\frac{1}{3}$ feet (40 inches)
Board D. $4\frac{1}{2}$ feet (54 inches)
Board E. $3\frac{1}{2}$ feet (42 inches)

4. a) LABEL: ounces (or cups)
b) ANSWER: **(4) B, E, A, C, D**

A. 12 ounces ($1\frac{1}{2}$ cups)
B. 4 ounces ($\frac{1}{2}$ cup)
C. 16 ounces (2 cups)
D. 24 ounces (3 cups)
E. 6 ounces ($\frac{3}{4}$ cup)

• • • • • • • •

Exercise 7
What to Do with Remainders
Page 123

1. a) Answer with remainder: 4 r 20
b) Round up
c) The school needs another bus to carry the remaining students.
d) Answer: 5

2. **a)** Answer with remainder: 5 r 1
   b) Round up
   c) She will need more than 5 pieces to make the dresses.
   d) Answer: 6

3. **a)** Answer with remainder: 4 r $\frac{1}{2}$
   b) Round down
   c) He cannot make more than 4 complete batches.
   d) Answer: 4

Mixed Review

Page 125

1. **(4) 100**
   Step 1: 100 + 50 + 150 + 150 + 100 + 50 = 600
   Step 2: 600 ÷ 6 months = 100

2. **(5) not enough information is given**
   The graph shows the number of inmates. It gives no information about drug or other offenses.

3. **(5) .69 (300)**
   percent
   repeat offenders
   ↓
   .69 (300) ← number of inmates in July

4. **(2) $8**
   Step 1: $14 \times \frac{1}{2}$ = 7 gallons
   Step 2: $1.17 per gallon × 7 gallons = $8.19
   Step 3: $8.19 ≈ $8

5. **(3) the hours he has already worked today**

6. **(2) 4,450 – .70 (4,450)**
   Step 1: 40% + 30% = 70%
   70% = .70
   Step 2: Total – 70% of total = undecided
   4,450 – .70 (4,450)

7. **(4) $59.99** – .30 ($59.99)
   original price discount
   ↓ ↓
   $59.99 – .30($59.99)

8. **(1) A**
   Cereal A: $1.68
   Cereal B: $1.74
   Cereal C: $1.98
   Cereal D: $1.69
   Cereal E. $2.25 (The $.35 applies to doughnuts.)

9. **(4) 35**
   Step 1: 25% = .25
   .25 × 28 ounces = 7 ounces
   Step 2: 28 + 7 = 35 ounces

10. **(3) 6**
   Step 1: 6 inches = $\frac{1}{2}$ foot
   Step 2: 1 yard = 3 feet
   Step 3: 3 feet ÷ $\frac{1}{2}$ foot =
   $3 \times \frac{2}{1} = 6$
   6 sections are needed to cover the border.

Chapter 7: Checking the Answer

Exercise 1

Did You Answer the Question?

Page 130

1. **a)** Correct solution: **(2) 1,150**
   200 + 750 + 90 + 110 = 1,150
   b) Choice (1) is incorrect because it answers the question: *How many total acres did the company have this year?*

2. **a)** Correct solution: **(3) $403.53**
   Step 1: $404.00 + 200.00 = $604.00
   Step 2: $37.50 + $18.97 + $144.00 = $200.47
   Step 3: $604 – $200.47 = $403.53
   b) Choice (1) is incorrect because it answers the question: *How much did Maureen deposit?*

3. **a)** Correct solution: **(2) 2,000**
   If 65 percent are white, then 35 percent are black and other minorities.
   $.35 \times x = 700$
   $x = 700 \div .35$
   $x = 2{,}000$
   b) Choice (3) is incorrect because it answers the question: *How many inmates are white?*

4. **a)** Correct solution: **(2) $63.75**
   .15 ($40 + $35) =
   .15 × $75 = $11.25
   $75 – $11.25 = $63.75
   b) Choice (5) is incorrect because it answers the question: *How much did Floyd save?*

5. **a)** Correct solution: **(1) $2\frac{1}{4}$**
   2 cups = $\frac{1}{2}$ quart
   $1 + \frac{3}{4} + \frac{1}{2} = 2\frac{1}{4}$ quarts
   b) Choice (3) is incorrect because: *It is the result of just adding the numbers and not changing cups to quarts.*

Exercise 2

Is the Answer Reasonable?

Page 133

1. a) Correct solution: **(4) 134**

$$\begin{array}{r}157\\-\ 23\\\hline 134\end{array}$$

b) Choice (2) is unreasonable because: *she should weigh less, not more, after her weight loss.*

2. a) Correct solution: **(4) 5**

Step 1: $8 + 6 + 3 + 8 + 5 + 3 + 2 = 35$ hours

Step 2: 35 hours $\div$ 7 days = 5 hours per day

b) Choice (2) is unreasonable because: *the average should not be equal to the total of the individual hours.*

3. a) Correct solution: **(5) 490**

Step 1: $4 \times 60 = 240$ miles

$5 \times 50 = 250$ miles

Step 2: $240 + 250 = 490$ miles

b) Choice (4) is unreasonable because: *25 miles is a very short distance for a car traveling near the speed limit for 9 hours.*

Exercise 3

Checking Your Computation

Page 136

		Check:	Corrected:	Recheck:
1.	$\begin{array}{r}185\\5\overline{)975}\end{array}$	$\begin{array}{r}185\\\times\ 5\\\hline 925\end{array}$	$\begin{array}{r}195\\5\overline{)975}\end{array}$	$\begin{array}{r}195\\\times\ 5\\\hline 975\end{array}$
2.	$\begin{array}{r}105\\\times\ 30\\\hline 31{,}500\end{array}$	$\begin{array}{r}1{,}050\\30\overline{)31{,}500}\end{array}$	$\begin{array}{r}105\\\times\ 30\\\hline 3{,}150\end{array}$	$\begin{array}{r}105\\30\overline{)\ 3{,}150}\end{array}$
3.	$\begin{array}{r}1{,}009\\-\ 437\\\hline 562\end{array}$	$\begin{array}{r}562\\+\ 437\\\hline 999\end{array}$	$\begin{array}{r}1{,}009\\-\ 437\\\hline 572\end{array}$	$\begin{array}{r}572\\+\ 437\\\hline 1{,}009\end{array}$
4.	$\begin{array}{r}481\\23\overline{)11{,}201}\end{array}$	$\begin{array}{r}481\\\times\ 23\\\hline 11{,}063\end{array}$	$\begin{array}{r}487\\23\overline{)11{,}201}\end{array}$	$\begin{array}{r}487\\\times\ 23\\\hline 11{,}201\end{array}$

Mixed Review

Page 137

1. (5) not enough information is given

You do not know how much each installment will be.

2. (3) 26

$640 \div 24 = 26.67$

The packer can fill only 26 cartons **completely.**

3. (3) Michigan

In 1987 Michigan's per-capita personal income was $15,000.

4. (5) 6 × 12/18

# of packages ↓   pencils per package ↓

$$\frac{6 \times 12}{18}$$

↑ # of students

5. (4) 8

hired → 3, turned away → 2

$$\frac{3}{2} = \frac{12}{x}$$

$3 \times x = 12 \times 2$

$3 \times x = 24$

$x = 24 \div 3$

$x = 8$

6. (2) 40

14-inch wire ÷ .35-inch pieces = 40 pieces

7. (2) $.75

Step 1: 2 × $1.50 = $3.00

Step 2: $\frac{1}{4}$ × $3.00 = $.75

8. (4) the amount each worker is paid per hour

total hours × $ per hour = total payroll

9. (3) $3.50

Step 1: Luxury dish soap = $1.60

Nature frozen corn ≈ $2.00

$1.60 + $2.00 = $3.60

Step 2: $3.60 ≈ $3.50

10. (3) $7.47

$2.49 per pound

× 3 pounds

$7.47

Chapter 8: Using Charts, Graphs and Drawings

• • • • • • • •

Exercise 1
"Reading" a Picture
Page 143

1. (4) 71
Find Wyoming on the graph and read across to the end of the bar.

2. (1) 20
Step 1: Find the percent value for Idaho—80%.
Step 2: Subtract 80 (percent outside metro area) from 100 (total percent) to find percent inside metro area:
100% – 80% = 20%

3. (3) .70 × 2,600,000

4. (5) $\frac{3}{4}$
Step 1: Find "rent" on the circle graph. The fraction is $\frac{1}{4}$.
Step 2: $1 - \frac{1}{4} = \frac{3}{4}$

5. (3) 3
Step 1: Add the fraction values for insurance and rent:
$\frac{1}{8} + \frac{1}{4} = \frac{1}{8} + \frac{2}{8} = \frac{3}{8}$
Step 2: Find the fraction value for supplies—$\frac{1}{8}$.
Step 3: $\frac{3}{8} \div \frac{1}{8} = 3$ or $\frac{3}{8}$ is *3 times* $\frac{1}{8}$

6. (4) $100,000
Step 1: Find the fraction value for insurance—$\frac{1}{8}$.
Step 2: $\frac{1}{8} \times \$800{,}000 = \$100{,}000$

• • • • • • • •

Exercise 2
"Reading Between the Lines" on Graphs
Page 146

1. (5) 20–21
Find the point where the line graph is at its lowest and look down to find the correct age group label.

2. (4) 91
Find the "14–15 years" label on the horizontal axis. Notice that it is a little above the 90 percent mark. This is approximately 91 percent.

3. (3) 3,025
Find the percent value for 18- and 19-year-olds. Notice that it is about halfway between 50 and 60, or 55. Find 55 percent of 5,500:
.55 × 5,500 = 3,025

4. (2) lose weight during their first week, then gain weight
Notice that all three lines go down between 0 and 1, then rise.

5. (2) 4
Step 1: Approximate the weight of a large baby at 6 weeks ≈ 11 pounds.
Approximate the weight of a small baby at 6 weeks ≈ 7 pounds.
Step 2: Subtract: 11 – 7 = 4

6. (1) 2
Step 1: Approximate the weight of an average baby at 1 week ≈ about 8 pounds.
Approximate the weight of a small baby at 1 week ≈ 6 pounds.
Step 2: Subtract: 8 – 6 = 2

• • • • • • • •

Exercise 3
Reading Graphs and Charts Carefully
Page 150

1. (1) 8
Step 1: fried beef liver—9 grams
Step 2: ground beef—17 grams

17 – 9 = 8

2. (3) 15
Step 1: 1 cup of 2 percent milk = 5 grams of fat
Step 2: 5 grams × 3 cups = 15 grams

3. (5) not enough information is given
The graph gives information about chicken without skin—you cannot tell how many grams of fat the skin contains.

4. (4) 162.175
Find the value listed under "MPH" for the category 1987.

5. (2) 2
1911: 6 hrs. 42 min. 8 sec. ≈ 7 hrs.
1971: 3 hrs. 10 min. 18 sec. ≈ 3 hrs.
7 is a little more than twice 3.

The 1911 time is a little more than *twice* the 1971 time.

6. (5) not enough information is given
The chart refers only to the *fastest* cars in the race—you are given no information about the slowest cars.

7. (2) 4
1987: 162.175 ≈ 162
1971: 157.735 ≈ −158
4

Mixed Review
Page 152

1. (5) not enough information is given
You are told that Hannah spends 5 hours relaxing and watching TV. You do not know how much of this time she spends in either activity.

2. (4) 7
Step 1: Hannah spends 3 hours cleaning and doing laundry. One-third of this time is equal to 1 hour.
Step 2: 6 hours (sleeping) + 1 hour = 7

3. (5) $\frac{1}{8}$
$\frac{\textit{hours spent preparing meals}}{\textit{total hours in a day}} = \frac{3}{24} = \frac{1}{8}$

4. (5) 25
$\frac{6 \text{ hours}}{24 \text{ hours}} = \frac{1}{4} = \frac{25}{100}$ OR 25%

5. (4) the number of people in Floyd's family
You know that each decoration takes 4 hours to make (1 + 2 + 1). You need to multiply this by the number of people in Floyd's family to find the total time.

6. (2) $2.88
$50.00 – $47.12 = $2.88

7. (1) $2.28 – $.20

8. (3) $3.36
Step 1: .05 × $3.20 = $.16
Step 2: $3.20 + $.16 = $3.36

Chapter 9: Working Geometry Word Problems

Exercise 1
Finding Information on Drawings
Page 156

1. (5) 142
Step 1: Bath: 8 ft × 4 = 32 ft
Family Room: (20 × 2) + (8 × 2)
40 + 16 = 56 ft
Kitchen: (15 × 2) + (12 × 2)
30 + 24 = 54 ft
Step 2: 32 ft + 56 ft + 54 ft = 142 ft

2. (4) 18
5 + 5 + 4 + 4 = 18

3. (5) not enough information is given
Although you can figure out the perimeter of the figure, you do not know the length of each brick to be used.

4. (4) $80
Area of 3 figures: (2 × 4) + (2 × 3) + (1 × 2)
8 + 6 + 2 = 16
Cost: $4.99 (≈$5.00) × 16 sq yd = ≈$80

5. (2) 3
1 foot + 1 foot + 1 foot = 3 feet

6. (5) not enough information is given
The window is a rectangle. You do not know the width of the rectangle.

Exercise 2
Let Formulas Work For You
Page 159

1. (2) 3 (3.14 × 10)
Circumference of a circle:
$C = \pi d$
$C = \pi \times 10$ in
$C = 3.14 \times 10$
Since the artist needs 3 figures, multiply the circumference by 3: 3 (3.14 × 10)

2. (4) 12

Perimeter of a triangle:

$P = a + b + c$

$= 12$ feet + 9 feet + 15 feet = 36 feet

$= 36$ feet ÷ 3 feet = 12 yards

(change feet to yards)

3. (3) 8,000

Volume of a cube:

$V = s^3$

$= 20 \times 20 \times 20 = 8{,}000$ cubic centimeters

4. (3) .50(8 × 12)

Area of a rectangle:

Area of floor $= lw$

$= 8 \times 12$

Cost: **$.50(8 × 12)**

Exercise 3

Visualizing Geometry Problems

Page 163

1. SPACE: area

CLUE: square miles

SOLUTION: $A = lw$

$= 4 \times 2.5 =$ **10 sq mi**

2. SPACE: volume

CLUE: cubic feet, container

SOLUTION: $V = s^3$

$= 2 \times 3 \times 2.5$

$= 15$ cu ft

= **67.5 pounds**

3. SPACE: perimeter

CLUE: braid to surround scarf

SOLUTION: $P = a + b + c$

$12 = 3 + 5 + c$

$12 = 8 + c$

$12 - 8 = c$

$\mathbf{c = 4}$ **feet**

Exercise 4

Picturing a Geometry Problem

Page 169

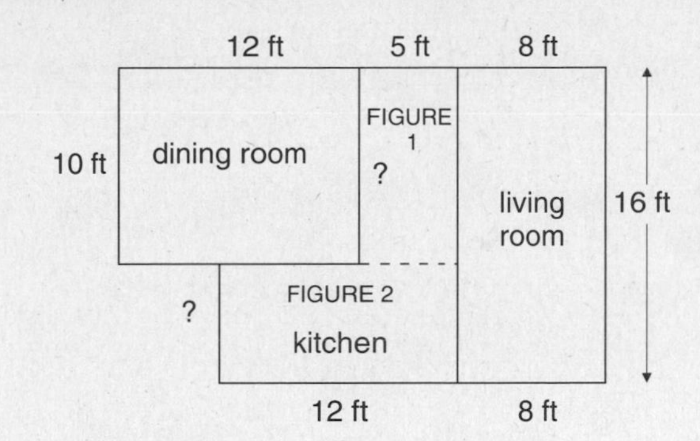

1. 28 sq yd

Step 1: living room

$A = lw$

$A = 16 \times 8 = 128$ sq ft

Step 2: dining room

$A = lw$

$A = 10 \times 12 = 120$ sq ft

Step 3: total area

128 sq ft + 120 sq ft = 248 sq ft

Step 4: change sq ft to sq yd

(9 sq ft = 1 sq yd)

$248 \div 9 = 27.6 \approx 28$ sq yd

2. $250.60

Step 1: Total area (see problem #1), area ≈ 28 sq yd

Step 2: $8.95 × 28 = $250.60

3. 122 sq ft

Step 1: Divide the kitchen into two figures (see the diagram above).

Step 2: Figure 1

$A = lw$

$A = 10 \times 5$ *(same length as dining room)*

$A = 50$ sq ft

Step 3: Figure 2

$A = lw$

$A = 6 \times 12$

16 ft (length of living room) − 10 ft (length of dining room) = 6 ft

$A = 72$ sq ft

Step 4: total area of kitchen

$A = 50 + 72 = 122$ sq ft

4. 200 miles

$c^2 = a^2 + b^2$
$c^2 = (160)^2 + (120)^2$
$c^2 = 25{,}600 + 14{,}400$
$c^2 = 40{,}000$
$c = \sqrt{40{,}000}$
$c = 200$

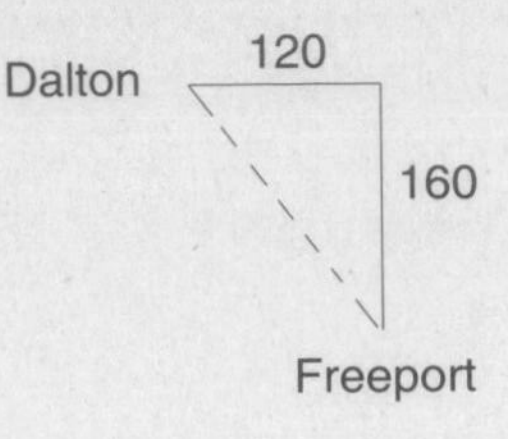

Mixed Review
Page 170

1. (4) 3.14×5^2

Area of a circle: $A = \pi r^2$
$A = 3.14 \times 5^2$

2. (2) $.008

1st bottle: $50\,\overline{)2.40}$ = .048

2nd bottle: $75\,\overline{)3.00}$ = .04

$$\begin{array}{r} .048 \\ -\ .040 \\ \hline .008 \end{array}$$

3. (5) 0

$$\begin{array}{r} \$\ 79.00 \\ +\ 49.00 \\ \hline 128.00 \end{array}$$

If Mary paid her bill of $128, she does not pay any interest.

4. (4) $100,000

Find 1989 on the horizontal axis and look up to the data line and across to the vertical value. It reads 100. Notice also that these values are in thousands.

5. (4) $10,000

1990 value: 100,000
1991 value: around 90,000

$$\begin{array}{r} 100{,}000 \\ -\ \ 90{,}000 \\ \hline 10{,}000 \end{array}$$

6. (2) 50,000

Find 1992 on the horizontal axis. The value for this year is projected to be about $50,000.

7. (2) $h = 57 \div (\frac{1}{2} \times 6)$

Area of a triangle: $A = \frac{1}{2}\,bh$
$57 = \frac{1}{2} \times 6 \times h$
$57 \div (\frac{1}{2} \times 6) = h$

8. (3) $558

Interest formula: $i = prt$
$\$44.64 = p \times .08 \times 1$
$\$44.64 \div .08 = p$
$\$558 = p$

9. (5) the radius of the can

Volume of a cylinder: $V = \pi r^2 h$
You know the value of π, and to find the height, you need to know the radius of the can.

10. (4) $\frac{1}{5}$

Total earnings: approximately $400 + $800 + $300 = $1,500
watercolor: approximately $300
$300 is $\frac{1}{5}$ of $1,500